Integrity Control
in
Parallel Database Systems

CIP-DATA KONINKLIJKE BIBLIOTHEEK, DEN HAAG

Grefen, Paul Willem Peter Johan

Integrity control in parallel database systems /
Paul Willem Peter Johan Grefen. - [S.l. : s.n.]. - Ill.
Thesis Enschede. - With index, ref.
ISBN 90-9005395-6
Subject headings: database systems.

Integrity Control in Parallel Database Systems

PROEFSCHRIFT

ter verkrijging van
de graad van doctor aan de Universiteit Twente,
op gezag van de rector magnificus,
prof.dr. TH. J. A. Popma,
volgens besluit van het College van Dekanen
in het openbaar te verdedigen
op vrijdag 16 oktober 1992 te 14.00 uur

door

Paul Willem Peter Johan Grefen

geboren op 18 april 1963
te Heerlen

Dit proefschrift is goedgekeurd door
Prof.dr. P. M. G. Apers, promotor
Dr.Ir. R. A. de By, referent

To my parents

Summary

Automated information systems have become an essential part of modern society. Industry, trade, and public services all heavily rely on the availability of large volumes of high-quality information. Nowadays, information is usually based on *databases*, large collections of data managed by specialized computer software referred to as *database management systems*. As the nature of information systems is changing, so are the requirements to database management systems. In the first place, the requirements to the quality of the data in databases are increasing. Consequently, modern database management systems must be equipped with powerful mechanisms, called *integrity control* systems, that guard the quality of data. In the second place, the sizes of databases are rapidly growing. This requires database systems that can manage and process large volumes of data in a fast way. The combination of these two developments results the requirement of integrity control systems that can both effectively and efficiently handle large databases.

This thesis combines a new approach to integrity control, referred to as *transaction modification*, with the techniques of *parallel database systems*. The transaction modification approach can effectively deal with a broad range of *integrity constraints*, which describe the quality requirements to a database. As suggested by its name, the technique is designed around the *transaction* concept. This guarantees both simple semantics and a number of other desirable properties of integrity control, like serializability and atomicity. The use of parallel database system techniques enables the efficient processing of large volumes of data in integrity control. Further, the use of parallelism allows the use of relatively simple and cheap computer architectures for powerful database systems including integrity control facilities.

The transaction modification technique is both discussed in conceptual terms and applied in a real-world parallel database system, called PRISMA/DB. The conceptual description of the technique is first developed in a sequential context, and next extended to parallel database systems. The application of the technique in PRISMA/DB clearly shows the feasibility of transaction modification. Measurements on this system demonstrate the positive effects of parallelism on the performance of integrity control. As transaction modification can lead to complex transactions, attention is paid to the scheduling of complex transactions in a parallel database system. Further, extensions to the transaction modification approach are discussed, leading in the directions of complex constraint types, active databases, and very high performance parallel integrity control systems.

Samenvatting

Geautomatiseerde informatie-systemen zijn een essentieel onderdeel geworden van de moderne samenleving. Industrie, handel en dienstverlening zijn in sterke mate afhankelijk van de beschikbaarheid van grote hoeveelheden informatie van hoge kwaliteit. Informatie is tegenwoordig veelal gebaseerd op gegevensbanken oftewel *databases*, grote verzamelingen gegevens die beheerd worden door gespecialiseerde computer-programma's, aangeduid met de term *data base management systems*. Door het veranderen van de aard van informatie-systemen veranderen ook de eisen die gesteld worden aan database management systems. Op de eerste plaats nemen de eisen toe met betrekking tot de kwaliteit van de gegevens in een database. Dientengevolge moeten moderne database management systems uitgerust worden met krachtige mechanismen voor het bewaken van de kwaliteit; deze mechanismen worden aangeduid als *integrity control systems*. Op de tweede plaats is een snelle toename te constateren van de omvang van databases. Deze ontwikkeling vereist database systems die grote hoeveelheden gegevens snel kunnen beheren en verwerken. De combinatie van de twee hierboven aangeduide ontwikkelingen leidt tot de vereiste van integrity control systems die zowel effectief als efficiënt kunnen omgaan met grote databases.

Dit proefschrift beschrijft de combinatie van een nieuwe aanpak van integrity control, *transaction modification*, en de technieken van *parallel database systems*. De transaction modification techniek kan op effectieve wijze om gaan met een breed scala van *integrity constraints*, die de beschrijving vormen van de kwaliteitsvereisten met betrekking tot een bepaalde database. Zoals gesuggereerd door de naam, is de techniek ontworpen rond het concept van *transacties*. Dit staat borg voor een eenvoudige semantiek en een aantal andere wenselijke eigenschappen van integrity control, zoals serialiseerbaarheid en atomiciteit. Het gebruik van technieken van parallel database systems maakt het met hoge snelheid verwerken mogelijk van grote hoeveelheden gegevens door integrity control systems. Verder kunnen door het gebruik van parallellisme relatief eenvoudige en goedkope computer-architecturen gebruikt worden voor database management inclusief integrity control.

De transaction modification techniek wordt in dit proefschrift zowel op conceptuele wijze beschreven alsook toegepast in een bestaand parallel database system, genaamd PRISMA/DB. De conceptuele beschrijving wordt eerst in een sequentiële omgeving ontwikkeld en vervolgens naar een parallelle omgeving uitgebreid. De toepassing van de techniek in PRISMA/DB toont duidelijk de realiseerbaarheid van transaction modification. Metingen aan dit systeem tonen aan, dat parallellisme een positieve invloed heeft op de prestatie van integrity control. Daar transaction modification kan leiden tot complexe transacties, wordt uitgebreid aandacht geschonken aan het in de tijd regelen van de executie van complexe transacties in parallel database systems. Tenslotte worden een aantal uitbreidingen van de transaction modification techniek besproken in de richtingen van complexe constraints, active databases en parallel integrity control systems met een zeer hoge prestatie.

Acknowledgements

Performing scientific research and writing down the results in a thesis is a task that cannot be performed well in solitude. Therefore, the credits for the work should not go to the researcher alone, but also to those people that have been an indispensable factor in the accomplishment, either in a professional or in a personal way. Here I want to express my thanks to these people.

My thesis advisor Peter Apers is thanked for offering the opportunity to perform the research the results of which are described in this thesis, and for his scientific advice during the past five years. Rolf de By is thanked for his role as thesis reviewer, in which he provided me with many and detailed comments on earlier versions of this thesis. Especially his help with the multi-set constructs in Chapter 2 has to be mentioned here. Further, his knowledge of the LaTeX text processing system has greatly contributed to the way the thesis looks now. Jan Flokstra has been an indispensable colleague in the practical aspects of the research. The work on integrity control in the PRISMA/DB context presented in Chapter 6 would not have been possible without him. Jennifer Widom from IBM Almaden Research Center at San Jose is thanked for her detailed comments on Chapters 1 to 5 of an earlier version of this thesis and her suggestions for improvements. All members of the PRISMA project from Philips Research Labs at Eindhoven, the Center for Mathematics and Computer Science at Amsterdam, the University of Amsterdam, and last but not least the University of Twente are thanked for providing a challenging research environment.

I want to thank my wife Ria for all her support and friendship during research and writing, even at times when I may have been rather bothersome. She always knew how to give me the right push when the speed in my work was getting low. Further, she is acknowledged for pointing out a number of 'personal' flaws in my use of the English language. My parents are thanked for providing the opportunity to start an academic career and for always supporting me.

Finally, all colleagues and friends not mentioned above that have contributed in one way or the other are thanked.

Contents

1 Introduction — 1
- 1.1 Topic of this thesis — 1
- 1.2 Related research — 3
- 1.3 Structure of this thesis — 3

2 Databases and integrity — 5
- 2.1 Relational database concepts — 5
- 2.2 Database integrity concepts — 14
- 2.3 Integrity control concepts — 17

3 Integrity constraints — 21
- 3.1 Integrity constraints and integrity rules — 21
- 3.2 Classification of integrity constraints — 22
- 3.3 Specification of integrity constraints — 25
- 3.4 Specification of integrity rules — 28

4 Integrity control — 31
- 4.1 An overview of integrity control — 31
- 4.2 Transaction modification — 35
- 4.3 Integrity specification — 38
- 4.4 Integrity rule optimization and translation — 42
- 4.5 Constraint enforcement — 49
- 4.6 Operational aspects — 53
- 4.7 Abstract system architecture — 56

5 Fragmentation and parallelism — 59
- 5.1 Distribution and parallelism — 59
- 5.2 Distributed database concepts — 61
- 5.3 Integrity rule fragmentation — 64
- 5.4 Integrity rule optimization and translation — 71
- 5.5 Constraint enforcement — 76

6	**Integrity control in PRISMA/DB**	**79**
	6.1 Introduction to the PRISMA project	79
	6.2 Introduction to PRISMA/DB	82
	6.3 Integrity control in PRISMA/DB	86
	6.4 Performance evaluation	90
	6.5 Conclusions	98
7	**Dynamic action scheduling**	**101**
	7.1 Scheduling in parallel database systems	101
	7.2 Basic notions	102
	7.3 Order dependency and transaction optimization	103
	7.4 Resource dependency and concurrency control	109
	7.5 Integrating both worlds	113
	7.6 Architectural issues	115
	7.7 Action scheduling in PRISMA/DB	119
	7.8 Conclusions	123
8	**Extending the ideas**	**125**
	8.1 Extended constraint types	126
	8.2 Active databases	129
	8.3 Constraint enforcement protocol extensions	131
9	**Conclusions**	**137**
	9.1 Transaction modification approach	137
	9.2 Parallelism	138
	9.3 Theory and practice	139
	9.4 Current status and future research	139
A	**Extended relational algebra summary**	**151**
B	**Example database**	**155**
	Curriculum Vitae	**157**

Chapter 1

Introduction

1.1 Topic of this thesis

Information has great value in modern society. Industry, trade, government, and public service organizations all rely heavily on the availability of information. Nowadays, information is usually based on data stored in computer systems. A collection of data in computer systems is referred to as a database, and the specialized computer software that manages the data and performs operations on it is referred to as a database management system. Database management systems form an essential component of modern automated information systems. As the nature of information systems is changing, so are the requirements to the database management systems on which they are built. Two important trends can be distinguished in these requirements.

The first trend to be observed is that the growing importance of information systems based on databases implies higher requirements on the quality of data in databases. Erroneous data in industrial or government databases can have negative effects on the quality of the decisions taken, and can lead to large economic disadvantages. Erroneous data in hospital databases or aircraft maintenance databases can even have catastrophic effects. Therefore, modern applications require mechanisms in database management systems that guard the quality of data in the databases. The importance of such mechanisms is convincingly illustrated in [BMJ65, Noble86], where striking anomalies in medical records are presented. The task of guarding the quality of data in databases is usually called integrity control, and implies controlling the quality of individual data items as well as the quality of possibly complex relationships between multiple data items.

The second trend to be observed is that demands on the performance of database management systems are rapidly growing. This has two main reasons: changes in the volume of data to be managed and in the nature of the operations to be performed on the data. The volume of data to be managed by the systems is becoming larger because of a number of developments. In the first place, information systems get a wider span: systems cover administrations of multinational companies or entire countries. In the second place, modern applications like engineering or geographic information systems require the storage of large quantities of very detailed information. Finally, multi-media information systems require the storage of space-consuming data, like entire documents, or even sound and pictures. The operations to be performed on data in databases are becoming more complex because applications require information of a high level of abstraction (e.g. management information systems) or require complex computations on the data (e.g. engineering and statistical databases). The fact that database applications are moving from batch to interactive environments with the requirement of low response times further increases the demands on the performance of database management systems.

A solution for the problems associated with the first trend described above is the development of integrity control subsystems, i.e. software components of database management systems that guard the quality of the data managed by the system. The quality of the data is described by explicit conditions to be satisfied by the database, called integrity constraints. In the integrity control research field, attention is mainly devoted to functionality of the integrity control subsystems, and much less to performance-related issues, although the performance of integrity control subsystems is generally considered one of the major problems impeding their general use. The answer of database research to the second trend described above is the development of high-performance database systems. An important way to achieve high performance in this context is the use of parallel architectures, in which a number of traditional computer systems are combined into one powerful system. Most research in this field concentrates on low level details of these systems that are related to high-performance processing of queries.

The work described in this thesis takes a new approach by integrally combining the ideas of integrity control subsystems and parallel data processing. The integrity control subsystems deal with requirements regarding the quality of data in databases, but require high performance processing capabilities. Parallel data processing techniques are used to obtain high performance data processing, both to accomodate the requirements of the integrity control subsystems and to deal with the general high demands on database management systems as mentioned above. This combination leads to the notion of *integrity control in parallel database systems*, the title of this thesis.

In developing new concepts and techniques, this thesis uses the relational data model as the data modelling context. This choice is based on the fact that database systems conforming to the relational data model are the current standard.

1.2 Related research

As mentioned in the section above, the work in this thesis combines two aspects of database management systems: the design of integrity control techniques and the use of parallelism to deal effectively with the requirement of high performance data processing. As both aspects are rarely combined in research until now, the work in this thesis has two separate areas of related research. These areas are sketched briefly below; more elaborate discussions are presented in Chapters 4 and 5.

The first area of related research is the field of integrity control in relational database management systems. Concepts in this field emerged after the definition of the relational data model in [Codd70]. Systems research started in the mid-seventies with the development of the first complete relational database management systems. Important systems were developed in the System R project at IBM San Jose Research Laboratory [Astr76, Cham81], and in the INGRES project at the University of California [Stone75, Stone86a]. These systems included support for rather limited integrity control. More recent projects that have paid attention to integrity control are the SABRE/Sabrina project at INRIA and MASI in France [Gard83, Simon84], the successor of the INGRES project called POSTGRES [Stone86b, Stone90a], and the Starburst project at IBM Almaden Research Center [Haas90, Ceri90b, Lohm91].

The second area of related research is the field of parallel database systems. Research in this field started in the mid-eighties, when performance limitations of traditional architectures became clear. Two directions in research have tried to deal with these problems, one developing special-purpose hardware for database machines, the other developing parallel architectures from standard components. The latter approach has received the most attention, because parallel architectures are relatively cheap, flexible, and scalable. Important projects in this context are the GAMMA project at the university of Wisconsin [DeWi90], the Bubba project at the MCC research center in Austin [Boral90], the HC16-186 project at the University of Trondheim [Brat89], and the PRISMA project [Apers92a, Apers92b] at various institutions in the Netherlands. The PRISMA project forms the system context for the research presented in this thesis.

1.3 Structure of this thesis

Chapters 2 and 3 describe the fundamental database concepts for the work in this thesis. Chapter 2 discusses the structures of and operations on relational databases, and the notion of integrity in the context of relational database systems. The integrity of a database is defined by means of explicit rules, called integrity constraints or integrity rules. These are dealt with in detail in Chapter 3.

Chapter 4 describes algorithms and abstract system architecture for integrity control in a centralized database system. The techniques described here are based on the *transaction modification* approach. As suggested by its name, this approach takes transactions as the basis for integrity control. Chapter 5 then extends the concepts of the previous chapter to obtain algorithms and abstract system architecture for integrity control in a distributed and parallel context.

In Chapter 6, the application of the abstract architecture mentioned above in a real-world parallel database system management system is discussed. In the work described here, the PRISMA/DB prototype database system is used as an implementation platform for the techniques developed in the previous chapters, and as a testbed for functionality and performance of these techniques.

Chapter 7 describes a scheduling technique for transaction processing in a parallel environment. This technique allows for parallelism between parts of complex transactions, thereby reducing the execution times of transactions. As transactions easily get complex in an application context with many integrity constraints, the scheduling technique is indispensable here to obtain a good performance.

Chapter 8 discusses a number of extensions to the transaction modification approach as discussed in Chapters 4 and 5. These extensions can be used to improve the functionality and performance of an integrity control subsystem.

Conclusions are presented in the last chapter of this thesis. The first appendix gives a short description of the database manipulation language used in this thesis, called *extended relational algebra* (XRA). The second appendix describes the example database that is used throughout this thesis to illustrate the various concepts and techniques.

Chapter 2

Databases and integrity

This chapter presents an introduction to the basic concepts of relational databases and integrity as used in this thesis. These concepts lay the foundation for the discussion of integrity constraints and integrity control in the next two chapters.

The first section below discusses the basic notions of relational databases in terms of the data structures and the operations defined on these structures. To accomodate the development of techniques in the sequel of this thesis, some deviations from and additions to the standard theory are made. The second section of this chapter introduces the concepts of database integrity and integrity constraints, and relates them to the notion of transactions. The last section pays attention to the basic ideas related to maintaining the integrity of the database. These ideas form the basis for the development of integrity control mechanisms in a database management system.

2.1 Relational database concepts

As stated above, the relational data model consists of *structures* and *operations*. This section first presents the *structures*, representing the static properties of the model; as explained below, integrity constraints are *not* considered to be part of the basic structures. After the structures, the *operations* on databases are discussed; the operations represent the dynamic properties of the relational data model. These concepts are then used for the description of *transactions*; the transaction concept plays a central role in this thesis.

2.1.1 Structures

The relational data structure concepts presented below are based on the standard theory as defined originally in [Codd70], and described thereafter in many textbooks like [Ullm82, Tsich82, Korth86, Elmas89]. Two main deviations from the standard approach can be distinguished here, however.

In the first place, the definition of the relational data model used in this thesis differs from the standard definition in the fact that relations are defined as multi-sets of tuples instead of sets of tuples. This choice is made because the use of multi-sets is closer to the practice of databases, and allows the definition of aggregate functions within the model.

In the second place, integrity constraints are treated separately from the database schemas in this thesis, although they are considered to be part of the schemas in many approaches [Codd79, Tsich82]. Integrity constraints play such a central role in this thesis that a separate and more elaborate treatment is justified.

Below, the basic concepts of relational database structures are defined. The first basic notion is that of a *domain*.

Definition 2.1 A *domain* $\mathcal{A}$ is a set of atomic values. The term *atomic* refers to the fact that each value in the domain is indivisible as far as operators of the relational data model are concerned. Each domain contains a special null value denoted as *null*, representing the situation where a value is unknown. □

Common types of domains are the basic data types of integers, reals, booleans, and strings. More specialized types as time, date, or money are possible too; note that they are also atomic in the sense of the definition above.

Definition 2.2 A *relation schema* $\mathcal{R}$ consists of a relation name and a list of attributes $\langle A_1, \cdots, A_n \rangle$. Each attribute A_i is defined on a domain $dom(A_i)$. The type of $\mathcal{R}$ is defined as $dom(\mathcal{R}) = dom(A_1) \times \cdots \times dom(A_n)$. A *relation* or *relation instance* R of relation schema $\mathcal{R}$ is a multi-set of elements in $dom(\mathcal{R})$, i.e. a function $R : dom(\mathcal{R}) \to I\!N$, where $I\!N$ denotes the domain of the natural numbers. The value of $R(x)$ is called the *multiplicity of x in R*. □

Definition 2.3 A *tuple* r of schema $\mathcal{R}$ is an element in $dom(\mathcal{R})$. A tuple r is an element of relation R if its multiplicity in R is greater than zero: $r \in R \Leftrightarrow R(r) > 0$. The value of the ith attribute of tuple r is denoted as $r.i$. The number of attributes of r is denoted as $\#r$. The projection $\pi_\alpha(r)$ is obtained by concatenating the attributes from r as specified by the attribute list α into a new tuple. In this, α is a list of prefixed integers $\langle \%i_1, \ldots, \%i_n \rangle$ with $n \geq 1$ and $1 \leq i_j \leq \#r$ for $1 \leq j \leq n$. The *concatenation* of two tuples $r_1 \oplus r_2$ is defined as the concatenation of the attributes of r_1 and r_2 in the specified order. The equality of two tuples $r_1 = r_2$ is defined for tuples having the same schema; $r_1 = r_2$ holds if all corresponding attributes of r_1 and r_2 have equal values. □

As stated above, relations are defined as multi-sets of tuples; this means that duplicate tuples are allowed in a relation. Multi-sets can be denoted as a collection of individual tuples r, possibly containing duplicates, or as a set of pairs $(r, R(r))$, without duplicates. Further, attributes in a relation schema are ordered to enable attribute addressing by index, rather than by name. This is a notational convention that implies no restrictions with respect to the situation with explicit attribute names, but enables addressing the attributes of anonymous relations. Attribute numbers are prefixed in attribute lists to avoid ambiguity with normal integer constants.

Definition 2.4 A *database schema* $\mathcal{D}$ is a set of relation schemas $\{\mathcal{R}_1, \cdots, \mathcal{R}_n\}$. A *database* or *database instance* D of database schema $\mathcal{D}$ is a set of relation instances $\{R_1, \cdots, R_n\}$. The set of all possible database instances of schema $\mathcal{D}$ is called the *database universe* $U_\mathcal{D}$, so $U_\mathcal{D} = dom(\mathcal{R}_1) \times \cdots \times dom(\mathcal{R}_n)$. □

Note that a database schema is a set of relation schemas; consequently, relations in a database are always addressed by name. A database instance is also commonly referred to as *database state*, this to clearly distinguish the concept from a database transition, as defined below.

Definition 2.5 A *database transition* of database schema $\mathcal{D}$ is an ordered pair of database states $\langle D^{t_1}, D^{t_2} \rangle$ of schema $\mathcal{D}$, with $t_1, t_2 \in I\!N$ and $t_1 < t_2$. The values t_1 and t_2 are called the *logical time* of the database states. □

Usually, a database transition describes two successive states of the database, so $t_2 = t_1 + 1$ in the definition above. This type of transition is called a *single-step transition*. If not stated otherwise, the term transition is used for single-step transitions in this thesis.

2.1.2 Operations

Below, basic operations on relational databases are introduced. First, the standard relational algebra is discussed. The constructs in this algebra are based on the relational algebra operators as they can be found in many textbooks on database systems, e.g. [Ullm82, Korth86, Elmas89]. Note, however, that they are modified to deal with multi-sets of tuples. The standard relational algebra allows the definition of expressions on relational databases. Next, this algebra is extended to obtain a complete language for the specification of programs to be executed against relational databases.

Standard relational algebra

Below, the basic relational algebra is defined. The algebra is then extended with some additional constructs that do not enhance the expressiveness of the algebra, but make life somewhat easier. Similar to the notation for relations, the multiplicity of a tuple x in a multi-set expression E is denoted as $E(x)$.

Definition 2.6 The *basic relational algebra* defines basic relational expressions

[Korth86]. A database relation is a basic relational expression. Let E_1, E_2, and E_3 denote basic relational expressions; E_1 and E_2 are defined on schema $\mathcal{E}$, E_3 is defined on schema $\mathcal{E}'$. Then the following constructs are basic relational expressions:

- The *union* $E_1 \cup E_2$ collects the elements of two multi-set expressions:
$$E_1 \cup E_2 = \{\, (x, E_1(x) + E_2(x)) \mid x \in dom(\mathcal{E}) \,\}$$

- The *difference* $E_1 - E2$ "subtracts" the contents of E_2 from the contents of E_1:
$$E_1 - E_2 = \{\, (x, max(0, E_1(x) - E_2(x))) \mid x \in dom(\mathcal{E}) \,\}$$

- The *product* $E_1 \times E_3$ forms the cartesian product of the elements of E_1 and E_3:
$$E_1 \times E_3 = \{\, (x \oplus y, E_1(x) \cdot E_3(y)) \mid x \in dom(\mathcal{E}) \wedge y \in dom(\mathcal{E}') \,\}$$

- The *selection* $\sigma_\varphi(E_1)$ selects elements from a multi-set that meet a condition φ defined on individual tuples in $dom(\mathcal{E})$:
$$\sigma_\varphi(E_1) = \{\, (x, E_1(x)) \mid x \in dom(\mathcal{E}) \wedge \varphi(x) \,\} \cup \{\, (x, 0) \mid x \in dom(\mathcal{E}) \wedge \neg\varphi(x) \,\}$$

In this definition, φ can be seen as a function from $dom(\mathcal{E})$ into the boolean domain.

- The *projection* $\pi_\alpha(E_1)$ projects a multi-set E_1 on the attributes in attribute list α[1]:
$$\pi_\alpha(E_1) = \{\, (\pi_\alpha(x), \Sigma_{\varphi(y)} E_1(y)) \mid x \in dom(\mathcal{E}) \,\}$$
where
$$\varphi(y) \equiv y \in dom(\mathcal{E}) \wedge \pi_\alpha(y) = \pi_\alpha(x)$$

$\square$

Note that attribute numbers in selection conditions and projection attribute lists are prefixed to avoid abiguity with normal integer constants.

Definition 2.7 The *standard relational algebra* is defined here as the basic relational algebra extended with two additional constructs. Any basic relational expression is a standard relational expression. Let E_1, E_2, and E_3 denote standard relational expressions; E_1 and E_2 are defined on schema $\mathcal{E}$, E_3 is defined on schema $\mathcal{E}'$. Then the following constructs are standard relational expressions:

---

[1] Here the summation $\Sigma_{\varphi(x)} f(x)$ is to be interpreted as the sum of $f(x)$ for all x satisfying $\varphi(x)$.

2.1 Relational database concepts

- The *intersection* $E_1 \cap E_2$ produces a multi-set consisting of the elements that are both in E_1 and E_2:

$$E_1 \cap E_2 = E_1 - (E_1 - E_2)$$

- The *join* $E_1 \bowtie_\varphi E_2$ produces a selection on the product of E_1 and E_2:

$$E_1 \bowtie_\varphi E_3 = \sigma_\varphi(E_1 \times E_3)$$

$\square$

The above definition of the intersection is equivalent to the following, perhaps more intuitive, definition:

$$E_1 \cap E_2 = \{\, (x, min(E_1(x), E_2(x))) \mid x \in dom(\mathcal{E}) \,\}$$

Extended relational algebra

The standard relational algebra above can be used for the specification of standard relational algebra expressions. The algebra lacks some important expressive power however: arithmetic expressions on attributes are not possible, duplicates cannot be removed, and aggregates over multi-sets are not included. Further, the algebra only allows the specification of single expressions, not of statements and programs to be executed against a database. The definition of the extended relational algebra below includes all these features. Note that the extended relational algebra is *not* an algebra in the mathematical meaning of the word.

Definition 2.8 The *multi-set aggregate functions* compute an aggregate value on a specified attribute of a multi-set expression. Let E be a multi-set defined on schema $\mathcal{E}$, and β an attribute of $\mathcal{E}$. The multi-set aggregate functions are defined as follows:

- The *count*: $CNT(E, \beta) = \sum_{x \in dom(\mathcal{E})} E(x)$
- The *sum*: $SUM(E, \beta) = \sum_{x \in dom(\mathcal{E})} x.\beta \cdot E(x)$
- The *average*: $AVG(E, \beta) = SUM(E, \beta)/CNT(E, \beta)$
- The *minimum*: $MIN(E, \beta) = min\{\, x.\beta \mid x \in dom(\mathcal{E}) \wedge E(x) > 0 \,\}$
- The *maximum*: $MAX(E, \beta) = max\{\, x.\beta \mid x \in dom(\mathcal{E}) \wedge E(x) > 0 \,\}$

The attribute parameter β in the CNT function is a dummy parameter, included only for reasons of syntactical uniformity. In the SUM and AVG functions, β must have a numeric domain. $\square$

Note that the set of aggregate functions defined above is rather arbitrary; other choices can be made, including statistical aggregate functions for example. Note further that the average, minimum and maximum functions are in fact partial functions, since they are not defined on empty multi-sets.

Definition 2.9 The *extended relational algebra expressions* are defined as the standard relational expressions extended with three additional constructs. Any standard relational expression is an extended relational expression. Let E be an extended relational expression defined on schema $\mathcal{E}$. Then the following constructs are extended relational expressions:

- The *extended projection* $\pi_\alpha(E)$ is similar to the normal projection defined above, but α contains arithmetic expressions defined on the attributes of E, rather than attributes of E only. These arithmetic expressions can be seen as functions from $dom(\mathcal{E})$ into a basic domain. Given $\alpha = \langle e_1, \ldots, e_n \rangle$ with $n \geq 1$, the extended projection on a tuple x is defined as:

$$\pi_\alpha(x) = [e_1(x), \ldots, e_n(x)]$$

Here, the square brackets denote tuple construction. Given this redefinition of the tuple projection, the definition of the extended projection operator on multi-sets is the same as the definition of the normal projection operator given before in Definition 2.6. The normal projection operator can be seen as a special case of the extended operator. The extended projection is denoted with the same symbol as the normal projection for reasons of readability; in the sequel of this thesis, the π symbol denotes the extended projection.

- The *unique* expression $unique(E)$ calculates the multi-set obtained by duplicate removal on E:

$$unique(E) = \left\{ \begin{array}{l} (x,1) \\ (x,0) \end{array} \middle| \begin{array}{l} x \in dom(E) \wedge E(x) > 0 \\ x \in dom(E) \wedge E(x) = 0 \end{array} \right\}$$

- The *groupby* expression $groupby(\alpha, f, \beta, E)$ on an expression E with schema $\langle A_1, \ldots, A_n \rangle$ calculates a multi-set aggregate function f on an attribute β producing a value in domain $\mathcal{F}$ per group of tuples, where the grouping is defined by equality of the attributes in the (duplicate-free) attribute list $\alpha = \langle \%a_1, \ldots, \%a_k \rangle$:

$$groupby(\alpha, f, \beta, E) = \{ (x,1) \mid x \in G \} \cup \{ (x,0) \mid x \in D' \wedge x \notin G \}$$
where
$$G = \{ x \in D' \mid (\exists y \in E)(x = \pi_\alpha y \oplus [f(\sigma_{\%a_1=x.1 \wedge \cdots \wedge \%a_k=x.k}(E), \beta)]) \}$$
and
$$D' = dom(A_{a_1}) \times \cdots \times dom(A_{a_k}) \times \mathcal{F}$$

2.1 Relational database concepts

If the attribute list α is empty, the groupby expression calculates an aggregate function over the attributes of all tuples in a multi-set; in this case, the result is one single-attribute tuple:

$$groupby(\langle\rangle, f, \beta, E) = [f(E, \beta)]$$

□

Definition 2.10 The *extended relational algebra statements* are defined as follows. Let R be a database relation, and E an extended relational expression of the same schema. Then the following constructs are extended relational algebra statements:

- The *insert* statement $insert(R, E)$ adds the elements of E to relation R:

  $$insert(R, E) \equiv (R \leftarrow R \cup E)$$

- The *delete* statement $delete(R, E)$ removes the elements of E from relation R:

  $$delete(R, E) \equiv (R \leftarrow R - E)$$

- The *update* statement $update(R, E, \alpha)$ modifies the elements in the intersection between R and E according to the attribute expression list α with the same schema as E:

  $$update(R, E, \alpha) \equiv (R \leftarrow (R - E) \cup \pi_\alpha(R \cap E))$$

  Note that π_α is an extended projection operator here.

- The *assignment* $R = E$ assigns the multi-set E to a new and implicitly defined relational variable R:

  $$(R = E) \equiv (R \leftarrow E)$$

- The *query statement* $?E$ sends the result of expression E as output to the user of the database system; the statement has no effect on the database.

In this definition, the symbol $\leftarrow$ denotes replacement. □

Extended relational algebra statements can be grouped into *programs* as defined below to specify more complex operations on a database.

Definition 2.11 The *extended relational algebra programs* are defined as follows. Let a be an extended relational algebra statement and p an extended relational algebra program. Then the following constructs are extended relational algebra programs:

- The *empty program* P_ε.
- The *single-statement program* a.
- The *multi-statement program* $p; a$.

□

A few simple operators on extended relational algebra programs are defined below.

Definition 2.12 Let p denote an extended relational algebra program $a_1; a_2; \ldots; a_n$. Then the functions *head* and *tail* are defined as follows:

$$head(p) = a_1$$
$$tail(p) = a_2; \ldots; a_n$$

□

Definition 2.13 The *program concatenation operator* $\oplus$ combines the statements of two extended relational algebra programs. The *concatenation* of two programs p_1 and p_2 is defined as the program consisting of the operations of p_1 and p_2 in their respective order:

$$p_1 = a_1; a_2; \ldots; a_m$$
$$p_2 = a_{m+1}; a_{m+2}; \ldots; a_n$$
$$p_1 \oplus p_2 = a_1; a_2; \ldots; a_m; a_{m+1}; a_{m+2}; \ldots; a_n$$

□

2.1.3 Transactions

Operations executed against a database are grouped into transactions to form database programs with certain characteristics. The transaction concept as defined below plays a central role in the sequel of this thesis.

Definition 2.14 A *transaction* T consists of an extended relational algebra program $a_1; \cdots; a_n$ enclosed in *transaction brackets*, to be executed against a database D:

$$T = (a_1; a_2; \cdots; a_n)$$

The parentheses denote the transaction brackets, respectively *begin* and *end*. During the execution of the actions a_i, the database is in a number of *intermediate states*. These states are not normal database states as they may contain temporary relations defined by assignment statements. If the logical time of D is t, then the state after the execution of action a_i is denoted as $D^{t.i}$; $D^{t.0} \equiv D^t$ denotes the state before the execution of a_1. The *end* bracket takes care of the transition from $D^{t.n}$ to a normal database state: if the transaction can *commit*, temporary

2.1 Relational database concepts

relations are removed from $D^{t.n}$ and the result $D^{t.n}\downarrow$ is installed as D^{t+1}; if the transaction must *abort*, D^t is installed as D^{t+1}. The states $D^{t.1}, \cdots, D^{t.n}$ have no semantics beyond the execution of T. The pre-transaction state D^t and post-transaction state D^{t+1} are visible to other transactions as well. This means that T is executed in isolation [Ceri84]. □

Informally, a transaction is a unit of work executed against a database state. Speaking more formally, a transaction T can be seen as an operator that transforms a database state D into another state $T(D)$ [Gray81], and can thus be associated with a single-step transition of a database:

$$D \xrightarrow{T} T(D)$$

According to the basic transaction model, the execution of a transaction T must satisfy the following properties [Ceri84]:

Atomicity The execution of T must always satisfy the atomicity property; this means that the effect of any execution of T on the initial database state D must be such that either the effects of T are completed fully, or D remains unchanged. So, if $T = (a_1, \ldots, a_n)$, the following must hold:

$$(T(D) = D^{t.n}\downarrow) \vee (T(D) = D)$$

Serializability The execution of a transaction T_1 must always be serializable with the execution of another transaction T_2 that is executed concurrently with it; this means that the effect of concurrently running transactions must be the same as the effect of some serial execution of these transactions. So, for two transactions T_1 and T_2 one of the following must hold:

$$D \xrightarrow{\|T_1, T_2\|} T_1(T_2(D))$$
$$D \xrightarrow{\|T_1, T_2\|} T_2(T_1(D))$$

Here $\| T_1, T_2 \|$ denotes the concurrent execution of T_1 and T_2. Note that the serializability property is a direct consequence of the isolation property defined above.

Durability Once T has committed, it must be guaranteed that the results of its actions will never be lost, independent of subsequent failures of any kind.

In the next section, these properties will be extended with a fourth property that defines the correctness of a transaction.

2.2 Database integrity concepts

This section introduces the basic integrity concepts for relational database systems. First, the general concept of integrity in database systems is introduced. Next, integrity constraints are discussed as a means to explicitly describe the integrity of a database. The definition of integrity constraints is then used to describe the properties that the execution of transactions should satisfy to maintain the integrity of a database.

2.2.1 The integrity concept

As stated in Chapter 1, in a database system the correctness or accuracy of data is of great importance. There are a number of ways in which incorrect data may occur in a database. The following disciplines in database technology try to prevent certain classes of errors [Eswa75, Hamm75, Date90b]:

Security Control deals with preventing users from accessing and modifying data in a database in unauthorized ways. The security control subsystem of a DBMS keeps record of the authorization of users to perform certain operations on certain data and checks this authorization upon database access [Date83, Elmas89].

Concurrency Control deals with the prevention of inconsistencies caused by concurrent access of multiple users or applications to a database. The concurrency control subsystem orchestrates the access to the database such that the serializability property of transactions is guaranteed. In most cases a locking or time-stamping technique is used [Date83, Korth86, Elmas89].

Reliability Control deals with the prevention of errors due to the malfunctioning of system hardware or software. The reliability control subsystem uses recovery techniques to reinstall a correct database after system crashes [Date83, Korth86], and techniques like replication of data to increase the availability of the system [Ceri84].

Integrity Control deals with the prevention of semantic errors made by users due to their carelessness or lack of knowledge; the integrity control subsystem uses integrity rules to verify the database and operations on the database.

This thesis is concerned with the integrity control discipline, and develops techniques to be used for an integrity control subsystem.

The term *integrity* as used throughout this thesis refers to the *correctness* or *validity* of the data in the database, as defined explicitly by means of integrity rules or *integrity constraints*[2]. This implies that neither the full correctness with respect to the part of the real world modeled by the database, nor the completeness of the

---

[2] Note that there is quite some confusion in terminology here: the terms *integrity*, *consistency*, *validity*, *correctness* etc. may be used differently by different authors.

2.2 Database integrity concepts

facts stored in the database is guaranteed in general; a detailed discussion of this topic can be found in [Motro89].

2.2.2 Databases and integrity constraints

As stated above, the integrity of a database is stated explicitly by means of integrity constraints, i.e. rules that define properties to be satisfied by the database. Below, the concept of *integrity constraint* is defined. In this definition, constraints are divided into *state constraints* that describe properties of database states, and *transition constraints* that describe properties of database transitions. This distinction is necessary to come to a correct definition of integrity constraints; a more detailed classification of constraints is discussed in Chapter 3.

Definition 2.15 Let $\mathcal{D}$ be a database schema. A *state constraint* I^s is a boolean function that is evaluated over a database state $D \in U_\mathcal{D}$:

$$I^s : U_\mathcal{D} \to bool$$

□

Definition 2.16 A *correct database state* $D \in U_\mathcal{D}$ satisfies each element of a set of state constraints $\mathcal{I}^s = \{I_1^s, \cdots, I_m^s\}$ defined on $\mathcal{D}$. The set of correct database states with schema $\mathcal{D}$ and constraint set $\mathcal{I}^s$ is denoted as:

$$U_\mathcal{D}^{\mathcal{I}^s} = \left\{ D \in U_\mathcal{D} \;\middle|\; \bigwedge_{i=1}^{i=m} I_i^s(D) \right\}$$

□

A state constraint describes the static properties of a database, i.e. the properties that a database should satisfy at one given moment.

Definition 2.17 Let $\mathcal{D}$ be a database schema. A *transition constraint* I^t is a boolean function that is evaluated over a database transition defined on $\mathcal{D}$:

$$I^t : U_\mathcal{D} \times U_\mathcal{D} \to bool$$

□

Definition 2.18 A *correct database transition* $\langle D_1, D_2 \rangle$ defined on schema $\mathcal{D}$ satisfies each element of a set of transition constraints $\mathcal{I}^t = \{I_1^t, \cdots, I_n^t\}$ defined on $\mathcal{D}$. The set of correct database transitions with schema $\mathcal{D}$ and constraint set $\mathcal{I}^t$ is denoted as:

$$V_\mathcal{D}^{\mathcal{I}^t} = \left\{ \langle D_1, D_2 \rangle \in U_\mathcal{D} \times U_\mathcal{D} \;\middle|\; \bigwedge_{i=1}^{i=n} I_i^t(D_1, D_2) \right\}$$

A transition constraint describes the correct transitions of a database; as such, it describes dynamic properties of a database. Therefore, transition constraints are also referred to as *dynamic constraints*.

2.2.3 Transactions and integrity constraints

The integrity of the database in terms of integrity constraints as defined above has its effect on the set of transactions that are allowed to be executed and successfully commited against a database.

The fact that each database state has to satisfy all state constraints means that the execution of a transaction T on a correct state D may never result in a database state that violates any constraint in $\mathcal{I}^s$; so the following should hold:

$$\left(\bigwedge_{i=1}^{i=m} I_i^s(D)\right) \Rightarrow \left(\bigwedge_{i=1}^{i=m} I_i^s(T(D))\right)$$

or in short notation:

$$\mathcal{I}^s(D) \Rightarrow \mathcal{I}^s(T(D))$$

The fact that each database transition has to satisfy all transition constraints means that given a correct database state, the execution of a transaction T may never imply a transition that violates any constraint in $\mathcal{I}^t$; so the following should always hold:

$$\left(\bigwedge_{i=1}^{i=m} I_i^s(D)\right) \Rightarrow \left(\bigwedge_{i=1}^{i=n} I_i^t(D, T(D))\right)$$

or again in short:

$$\mathcal{I}^s(D) \Rightarrow \mathcal{I}^t(D, T(D))$$

Given these obeservations, we can define the correctness of a transaction as follows.

Definition 2.19 A transaction T is *correct* with respect to a correct database state D and a set of integrity constraints $\mathcal{I}$ if and only if a commited execution of T on D does not imply a database transition that violates any transition constraint in $\mathcal{I}$, and the post-transaction database state $T(D)$ does not violate any state constraint in $\mathcal{I}$. A transaction that does not comply with this requirement is called *incorrect*. □

The concept of *correctness* of a transaction is essentially different from the concept of *safeness* of a transaction as defined below.

Definition 2.20 A transaction T is *safe* with respect to a database schema $\mathcal{D}$ and a set of integrity constraints $\mathcal{I}$ if and only if a commited execution of T on any correct state $D \in U_\mathcal{D}$ does not imply a database transition that violates any transition constraint in $\mathcal{I}$, and the post-transaction database state $T(D)$ does not violate any state constraint in $\mathcal{I}$. A transaction that is not safe is called *unsafe*.$\square$

2.3 Integrity control concepts

The previous section has discussed the conceptual notion of integrity of a database. This section describes the fundamental issues regarding integrity control in a database system. First, the choices with respect to the responsibility for integrity control are given and compared; various observations lead to the approach taken in this thesis, in which this responsibility is completely with the database management system. The next question to be answered is how to couple the integrity control task to the other tasks of the system; the answer to this question sets the scene for the architectural issues of integrity control as discussed in Chapter 4. Finally, the fundamental concepts of handling integrity violations are discussed.

2.3.1 The responsibility for integrity control

If the integrity of a database has to be maintained, an important question to be answered is who or what is responsible for integrity control. In general, there are three main options for the allocation of the integrity control task:

Application Designer The integrity control task can be left to the application designer or ad hoc user of the database system. As such, no robustness with respect to integrity is offered at all, and applications are required to be safe.

Transaction Designer The integrity control task can be the responsibility of the transaction designer. This means that transactions are required to be safe. In this situation, applications can only make use of predefined transactions.

Database Management System The database management system (DBMS) can have the responsibility for integrity control. This means that arbitrary transactions can be executed on the system.

In the first case, the integrity of the database is not guaranteed by a central specification of the database itself, but only by the design of the (numerous and ever changing) applications operating on the database. Consequently, the chance of improper constraint control is high. Further, changes of some constraint definitions require modification of all applications including integrity control with respect to these definitions. This is an undesirable situation, although it may occur frequently in practice.

In the second case, the responsibility for integrity control is part of the transaction design process [Gard79]. Again, changes to constraint definitions require

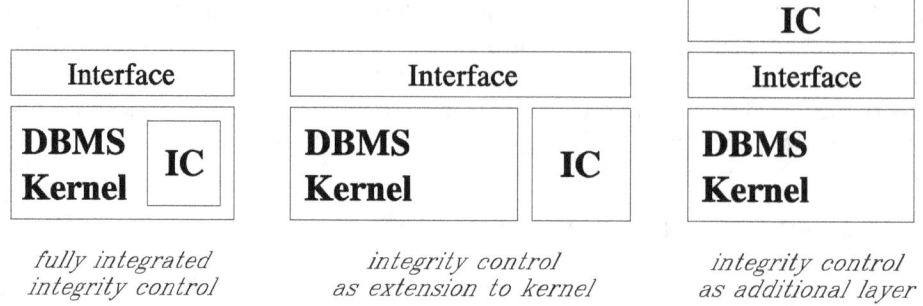

Figure 2.1: Coupling integrity control to a DBMS

modification of all transactions including integrity control with respect to these definitions. In a database application including many constraints, transactions can get very complex; a transaction design tool may be required here to alleviate the task of the transaction designer in this situation [Wang91]. Further, this approach is hardly feasible in a situation with a large number of ad hoc transactions.

If the integrity control task is allocated with the database management system, the integrity of the database is effectively ruled by a central set of constraint definitions managed by the system. Applications and transactions can be fully unaware of integrity control. A disadvantage of this approach may be a reduced flexibility with respect to constraint handling, because variations with respect to handling constraint violations per application or transaction type are not possible.

The latter two approaches are described in literature, and the choice between them depends on many criteria like the application domain, the actual use of a database system, performance requirements etc. The approaches are not necessarily mutually exclusive, however; they can be combined in practical situations, where multiple approaches are used for various classes of constraints. This thesis is mainly concerned with fully automatic integrity control with transparency to all users of the database system, and therefore adopts the DBMS-based approach.

Having allocated the responsibility for integrity control to the database management system, there are various ways to couple the integrity control subsystem to the system kernel, as depicted in Figure 2.1. One approach is to leave the system as it is, but add a layer on top of the system that takes care of integrity control; this approach is taken in the AIM project [Crem83]. A second approach is to have an extensible system kernel, and add integrity control as an extension to this kernel; this approach is taken in the Starburst project [Widom91]. A third approach is to see integrity control as part of the basic functionality of a database system kernel, and have a fully integrated integrity control subsystem; this approach is taken in most current database systems that deal with integrity control. This thesis also adopts the latter approach, because the database system kernel is seen as a transaction processing system, and integrity control is considered to be a base function of transaction processing. This choice is mainly based on a system engineering point of view, as all three approaches may result in the same external functionality.

2.3.2 Integrity violation handling

As discussed above, the approach to integrity control as described in this thesis assigns the responsibility for this task to the database management system. Therefore, the system must be able to deal with incorrect transactions effectively. The term 'effectively' has two aspects here:

- In the first place, the integrity control mechanism must in all cases prevent inconsistent states or transitions of the database. This means that so-called *soft assertions* [Eswa75] are not considered here, since this approach to integrity control only signals integrity violations, and leaves solving the integrity problem to the user.

- In the second place, dealing with an incorrect transaction T must be performed completely within the transaction boundaries of T, this to avoid violating the transaction atomicity principle. This means that the integrity violation handling mechanism must be fully integrated into the transaction execution mechanism.

Within the boundaries of the two conditions above, two approaches for dealing with incorrect transactions can be distinguished:

Aborting approach The aborting approach aborts the execution of an incorrect transaction T such that the pre-transaction database state is reinstated and no database transition occurs.

Compensating approach The compensating approach compensates the actions of an incorrect transaction T such, that a correct transaction T' is obtained that can be executed without the risk of integrity violation.

The aborting approach only requires the possibility to detect integrity violations, and relies further on general transaction execution mechanisms. In general, the generation of compensating actions for a given arbitrary transaction is of the same complexity as the design of correct transactions. Consequently, the aborting approach is clearly the simpler one. The aborting approach, however, will not always result in the functionality of integrity control as required by certain applications. Therefore, a complete integrity control subsystem should offer both approaches.

The chapters that follow describe concepts and mechanisms that can be used for both approaches. The emphasis is, however, on the aborting approach. This has two reasons. In the first place, the aborting approach is the most general, since it results in clear and simple semantics of integrity control for all types of constraints, whereas the compensating approach is hard to use for certain classes of constraints. In the second place, the emphasis on the aborting approach is of a pragmatic nature, since it allows for brevity and clearness in the presented techniques.

Chapter 3

Integrity constraints

This chapter discusses the concept of integrity constraints in detail. The first section describes the two appearances of integrity constraints: the purely declarative form that describes the conditions that should be satisfied by the database, and the operational form that includes information of how to enforce the constraints. The declarative form will be referred to simply as *integrity constraint*, and the operational form will be called *integrity rule*. Next, a classification of integrity constraints is discussed that is used to determine the constraint types discussed in this thesis. For these constraint types a specification formalism is developed for both the functional and the operational form in the last two sections of this chapter.

3.1 Integrity constraints and integrity rules

In the previous chapter, the notion of *integrity constraints* was defined. Given a database schema $\mathcal{D} = <\mathcal{R}_1, \cdots, \mathcal{R}_n>$, $U_\mathcal{D}$ is the set of all database states of $\mathcal{D}$. A state constraint I^s is a boolean function that is evaluated over a database state D in the database universe $U_\mathcal{D}$. A transition constraint I^t is a boolean function that is evaluated over a pair of database states $\langle D_1, D_2 \rangle$ in $U_\mathcal{D} \times U_\mathcal{D}$.

According to these definitions, an integrity constraint specifies a condition that should be satisfied by the database. It does not specify, however, when the constraint should be evaluated over the database, and what action is to be performed if an integrity violation is detected. If integrity constraints are to be used in a real world database system, this information is necessary for the integrity control task. Therefore, a more operational form of constraints is used, called *integrity rules*.

The specification of an integrity rule J includes a specification of the events on which the rule should be activated, the condition that must be met by the database, and the action to be taken if the condition is not satisfied.

Definition 3.1 An *integrity rule* defined on a database schema $\mathcal{D}$ is a triple $J = \langle t, c, a \rangle$ with the following elements [Gref90d]:

- The *trigger set* t is a set of pairs $\langle u, r \rangle$ with $u \in \{INS, DEL\}$ and r a relation name in $\mathcal{D}$; the trigger set specifies the update operations that may violate the constraint.

- The *condition* c is a declarative specification of the integrity constraint associated with the integrity rule; it describes the condition that should be met by the database in the formalism used for integrity constraints.

- The *violation response action* a is an extended relational algebra program as defined in Chapter 2; the violation response action specifies the actions to be taken if the condition c of the rule is not satisfied by the database.

If $J = \langle t, c, a \rangle$ denotes an integrity rule, $triggers(J)$ denotes the trigger set t, $condition(J)$ denotes the condition c, and $action(J)$ denotes the action a of J. □

As will be clear, a trigger $\langle INS, R \rangle$ is used for an insert operation on R and $\langle DEL, R \rangle$ for a delete operation. An update trigger is modeled as a combination of a delete and an insert trigger, i.e. an update operation triggers both rules with an insert trigger and rules with a delete trigger on the involved relation.

The relation between integrity constraints and integrity rules is one-to-many, since multiple rules exist with the same condition, but different trigger sets and violation response actions. In practice, each combination of a constraint and a trigger set is associated with a single action, since the violation of a constraint should not have an ambiguous effect. The most common violation response action will be the transaction abort, because other semantically meaningful actions may be hard to design for many constraints.

3.2 Classification of integrity constraints

The range of constraint types one can think of is unlimited, and many types and variations on them have been described in database literature. To bring some order in the wilderness, this section presents a simple classification of integrity constraints. This classification is used to first delimit the constraint types discussed in this thesis, and classify them thereafter. The basis for this is a number of classification criteria for constraints developed below.

3.2.1 Characterizing constraints

The ways to characterize integrity constraints are as many-fold as the constraint types one can envision. For the goals mentioned above, a number of simple classification criteria will be sufficient, however; these are presented below.

3.2 Classification of integrity constraints

Following the ideas in literature, integrity constraints can be classified based on the following orthogonal characteristics:

- The *space scope* of a constraint is the 'part' of the database the constraint is defined on. Usually, a distinction is made between attribute constraints, tuple constraints, relation constraints, and database constraints.

- The *time scope* of a constraint is the number of database states the constraint is defined on. An important distinction can be made between static constraints, single-step transition constraints, and general transition constraints.

- The *definiteness* of a constraint specifies if the constraint defines a 'hard' condition or a 'fuzzy' condition on the database [Raju88]. Fuzzy conditions are stated in some kind of fuzzy logic.

Besides the classification above, two other criteria are used commonly, which do not have an orthogonal relationship with each other or those above:

- The *nature with respect to the data model* specifies whether the constraint is considered an integral part of the data model or not. The first type of constraint is called an inherent or structural constraint, the second type an explicit constraint [Tsich82].

- The *computation complexity* of a constraint specifies the complexity of the algorithm needed for evaluation of the constraint. The evaluation can be linear, quadratic or exponential in terms of the sizes of the involved relations, for example.

Since the main topic of this thesis is not integrity constraint functionality, but constraint handling in a parallel context, the constraint types discussed in the sequel of this paper are limited to the types that are generally used in theory and practice; extensions with respect to these constraint types are discussed in Chapter 8. Constraints with all kinds of space scope are of general interest, so no restriction is made with respect to this characteristic. Both static and single-step transition constraints are used in many applications and well-understood in theory; general dynamic constraints [Ehri84, Chom92] are considered a difficult research topic, however. Therefore, the time scope of constraints is restricted to static and single-step dynamic. Since fuzzy constraints are a rather 'exotic' research topic, the definiteness of constraints is restricted to 'hard' constraints.

The distinction between inherent and explicit constraints is only important from a data modeling perspective, and will therefore not be made in this thesis (except for selecting important constraint types). The computation complexity of constraints will be used where necessary.

time scope	space scope	constraint type
state	attribute	domain
		nonull
	tuple	attribute comparison
	relation	uniqueness (key)
		functional dependency
		aggregate
	database	referential integrity
		interrelation aggregate
transition	attribute	domain
	tuple	attribute comparison
	relation	aggregate
	database	interrelation aggregate

Table 3.1: Constraint Taxonomy

3.2.2 Classifying constraints

In the section above, the constraint types discussed in this thesis have been delimited by making certain choices for their characteristics. Within these boundaries, the forms of constraints one can describe are many-fold. Therefore, it is very hard, if not impossible, to come to a detailed classification of all considered constraint types. Therefore, further choices are made to obtain a number of constraint types that will be used in the sequel of this thesis to illustrate the developed constraint handling techniques. The constraint types to be discussed in the sequel of this thesis at least include the following categories:

1. the inherent or structural constraint types of the relational model [Gard89, Elmas89]: unique key constraint, referential integrity constraint, domain constraint, and nonull constraint;

2. some other common state constraints: functional dependency and attribute comparison constraints.

3. transition constraints with attribute, tuple, relation, and database space scope;

4. aggregate constraints, both with one and multiple aggregate operators.

These requirements have been used to obtain a simple and well-organized set of constraint types. The resulting taxonomy of constraint types is shown in Table 3.1. Note that this taxonomy can easily be extended with further constraint types.

3.3 Specification of integrity constraints

For the specification of integrity constraints a constraint specification language has to be chosen. In the literature, several types of languages are used for this purpose: first order logic [Nico78, Small86], languages based on relational calculus [Gard89], derivations from query languages like SQL [Astr76, Ceri90a] and QUEL [Stone75], languages derived from logic programming languages [Noble86, Ceri90c], and various special purpose languages for constraint specification [Morg84, Gard89, Vald89].

The constraint specification language to be used in this thesis should be simple and generally accepted, should have a clear and formal semantics, should correctly deal with multi-sets, and should have enough expressive power to describe the constraint types listed in Section 3.2.2. A specification language based on the tuple relational calculus [Ullm82] is chosen, enhanced with some extensions based on the work in [By91].

3.3.1 Specification language

In this thesis, integrity constraints are specified as well-formed formulas in the integrity constraint specification language $\mathcal{CL}$. This language is intentionally kept simple, but powerful enough to express at least all the constraint types selected in the previous section; extension can be made easily.

The language $\mathcal{CL}$ is defined below in a number of steps. The first step is the definition of the alphabet of $\mathcal{CL}$, which provides the basic elements of the language.

Definition 3.2 The *alphabet* for the specification of integrity constraints consists of the following symbols (note that the {} brackets, commas, and ... are meta-symbols here):

- The set of value constants $C = \{c_1, c_2, \ldots\}$.

- The set of multi-set constants $M = \{R, S, R_1, R_2, \ldots\}$.

- The set of tuple variables $V = \{x, y, z, x_1, x_2, \ldots\}$

- The set of tuple function symbols $FT = \{.\}$ of type $V \times C \to C$, the set of value function symbols $FV = \{+, -, *, /\}$ of type $C \times C \to C$, the set of aggregate function symbols $FA = \{SUM, AVG, MIN, MAX\}$ of type $M \times C \to C$, and the set of counting function symbols $FC = \{CNT, MLT\}$ of type $M \to C$.

- The set of value predicate symbols: $PV = \{<, \leq, =, \neq, \geq, >\}$ of type $C \times C$, the set of multi-set predicate symbols $PM = \{\in\}$ of type $V \times M$, and the set of tuple predicate symbols $PT = \{=\}$ of type $V \times V$.

- The sets of unary connectives $CU = \{\neg\}$ and binary connectives $CB = \{\vee, \wedge, \Rightarrow\}$.

- The set of quantifiers: $Q = \{\exists, \forall\}$.
- The set of punctuation symbols: $IP = \{(,)\}$.

□

The multi-set constants from the set M correspond with the base and auxiliary relations of a database. The base relations are the permanent relations containing the data actually stored in the database. The auxiliary relations are calculated from the base relations automatically by the database management system for specific integrity control purposes. An important type of auxiliary relation is the pre-transaction state of a relation, necessary for the specification of transition constraints; an example follows.

The semantics of the function and predicate symbols in $\mathcal{CL}$ are assumed to be clear after the definitions in Chapter 2, except for the ones described here. The CNT function symbol from the set FC denotes the count or cardinality function, producing the number of elements in a multi-set. The MLT function symbol denotes the maximum multiplicity function; for a multi-set E defined on schema $\mathcal{E}$ this function is defined as follows:

$$MLT(E) = max\{E(x) \mid x \in \mathcal{E}\}$$

The function and predicate symbols presented above provide a 'basic collection'. The sets can be extended easily for specific situations. For example, in a statistical context, one may want to include the standard deviation and variance function symbols in FA.

Definition 3.3 The elements of the set of *terms* $\mathcal{T}$ are the following:

- A value constant from the set C.
- An attribute selection $x.i$, where $x \in V$ and i an integer constant from C.
- An arithmetic function application $t_1 \vartheta t_2$, where $\vartheta \in FV$, $t_1 \in \mathcal{T}$, and $t_2 \in \mathcal{T}$.
- An aggregate function application $\Gamma(R, i)$, with $\Gamma \in FA$, $R \in M$, and i an integer constant from C.
- A counting function application $\Gamma(R)$, with $\Gamma \in FC$ and $R \in M$.

□

As defined above, attribute selection $x.i$ on a tuple x is performed by specifying the index i of the attribute within the tuple. If explicit attribute names are available, attribute selection by specifying the attribute name is also allowed to enhance readability. In this case, the attribute name can be seen as an alias for the corresponding attribute index.

3.3 Specification of integrity constraints

The terms of the language $\mathcal{CL}$ are used to construct the atomic formulas of the language as defined below.

Definition 3.4 The elements of the set of *atomic formulas* $\mathcal{A}$ are the following:

- An arithmetic comparison $T_1 \vartheta T_2$, with $\vartheta \in PV$, and $T_1, T_2 \in \mathcal{T}$.
- A multi-set membership expression $x \in R$, where $x \in V$, and $R \in M$.
- A tuple comparison $x = y$, where $x \in V$ and $y \in V$; note that the tuple comparison tests the equality of two tuples, i.e. the equality of the values of all corresponding attributes of x and y.

□

Finally, the atomic formulas are used for the construction of the well-formed formulas of the language, which can be used to specify integrity constraints.

Definition 3.5 The elements of the set of *well-formed formulas* $\mathcal{W}$ are the following:

- An atomic formula $\mathcal{A}$.
- A negation $\neg W$, with $W \in \mathcal{W}$.
- A connection $W_1 \vartheta W_2$, with $W_1, W_2 \in \mathcal{W}$, and $\vartheta \in CB$.
- A quantification $(\vartheta x)W$, with $\vartheta \in Q$, $x \in V$, and $W \in \mathcal{W}$.

□

Finally, a few shorthand notations are introduced to enable constraint definitions that are easier to read:

- $(\forall x, y)W$ is short for $(\forall x)(\forall y)W$.
- $x, y \in R$ is short for $x \in R \land y \in R$.

3.3.2 Example constraints

Below, a few example constraints in the $\mathcal{CL}$ language as defined above are presented. The constraints are defined on the example beer database discussed in Appendix B; the constraint names refer to those in this appendix. For the convenience of the reader, an overview of the relations in the example database is given in Table 3.2.

The first example constraint $I1$ is a simple domain constraint on relation *beer*, stating that a beer cannot contain a negative percentage of alcohol:

$(\forall x)(x \in beer \Rightarrow x.alcperc \geq 0)$

relation	attributes
beer	name, brewery, type, alcperc
brewery	name, city, country, part_of
pub	name, city, country, brewery
trade	pub, beer, qty_bought, qty_sold

Table 3.2: Relations of example database

The next example $I2$ is a uniqueness (key) constraint, stating that the name of a brewery should be unique:

$(MLT(brewery) \leq 1) \wedge$
$(\forall x, y)((x, y \in brewery \wedge x.name = y.name) \Rightarrow (x = y))$

A referential integrity constraint $I3$ from relation *beer* to relation *brewery* is shown below; it states that every beer should be brewed by an existing brewery:

$(\forall x)(x \in beer \Rightarrow (\exists y)(y \in brewery \wedge x.brewery = y.name))$

The example below is an aggregate constraint $I6$ stating that the average alcohol percentages of all beers should exceed a certain value:

$AVG(beer, alcperc) > 3$

Finally, an example transition constraint $I7$ is presented. Suppose that a new version of an existing beer should always be at least as strong as the old version (which may be considered good brewing practice); then we have the following constraint:

$(\forall x, y)((x \in beer \wedge y \in beer_{old} \wedge x.name = y.name) \Rightarrow$
$(x.alcperc \geq y.alcperc))$

In this constraint, $beer_{old}$ is an auxiliary relation containing the pre-transaction state of relation *beer*.

3.4 Specification of integrity rules

For the specification of integrity rules a rule specification language must be chosen. Most approaches in current database systems are based on extensions of query languages, like SQL in the Starburst project [Widom90, Widom91], and QUEL in the POSTGRES project [Stone88, Stone90a]. Here an approach is chosen that makes use of the languages already introduced in the preceding chapters: the integrity constraint specification language $\mathcal{CL}$ is used for the specification of rule conditions, and the extended relational algebra as defined in Chapter 2 is used for the specification of rule actions.

3.4 Specification of integrity rules

3.4.1 Specification language

The integrity rule specification language $\mathcal{RL}$ is used for the specification of integrity rules. The language provides a user-friendly syntax for the rule construct defined in the first section of this chapter.

As mentioned above, specification languages for rule conditions and rule actions are already available. A specification language for trigger sets of rules is defined below.

Definition 3.6 The set of *trigger specifications* on database schema $\mathcal{D}$ is defined as follows. Let U denote an elementary update type, such that $U \in \{INS, DEL\}$, and let R denote a relation name in $\mathcal{D}$. Then the set of trigger specifications on $\mathcal{D}$ consists of all possible combinations $U(R)$. □

Trigger specifications can be combined into lists of trigger specifications, according to the following definition:

Definition 3.7 The set of *trigger set specifications* on relation schema $\mathcal{D}$ is defined as follows. Let t be a trigger specification on schema $\mathcal{D}$ and s a trigger set specification on $\mathcal{D}$. Then the following lists are trigger set specifications:

- The single element trigger set specification t.

- The composed trigger set specification consisting of the concatenation of a trigger and a trigger set specification s, t.

□

Now all ingredients for integrity rule specification are available. All that remains is to define how the ingredients are to be combined.

Definition 3.8 The set of integrity rule specifications in language $\mathcal{RL}$ is defined as follows. Let ts be a trigger set specification, c an integrity constraint specificationin $\mathcal{CL}$, and p an extended relational algebra program. Then the following construct is an integrity rule specification:

WHEN ts
IF NOT c
THEN p

□

3.4.2 Example rules

Below, a few example integrity rules in the language $\mathcal{RL}$ are given. The rules are defined on the example database presented in Appendix B and are derived from the example integrity constraints presented above. Efficiency is not yet an issue in these rules; optimization aspects follow in the next chapters.

The first example below is an integrity rule based on domain constraint $I1$ from the example database in Appendix B. The trigger set of this rule states that the rule should be triggered whenever new values have been added to the beer relation; clearly, a domain constraint cannot be violated by deleting values from a relation. The rule condition is the domain constraint associated with the rule. The violation response action specifies what is to be done to obtain a correct database; the rule below aborts an incorrect transaction.

> WHEN $INS(beer)$
> IF NOT $(\forall x)(x \in beer \Rightarrow x.alcperc \geq 0)$
> THEN $abort$

The rule below has the same constraint $I1$ as its condition, but a different violation response action. This clearly illustrates the one-to-many relationship between integrity constraints and integrity rules.

> WHEN $INS(beer)$
> IF NOT $(\forall x)(x \in beer \Rightarrow x.alcperc \geq 0)$
> THEN $delete(beer, \sigma_{alcperc<0}(beer))$

The integrity rule below is based on referential integrity constraint $I3$. The action of this rule inserts default tuples into the *brewery* relation for the beers that 'have no brewery'.

> WHEN $INS(beer), DEL(brewer)$
> IF NOT $(\forall x)(x \in beer \Rightarrow (\exists y)(y \in brewer \land x.brewer = y.name))$
> THEN $temp = unique(\pi_{\langle brewery \rangle} beer) - \pi_{\langle name \rangle} brewery;$
> $\qquad insert(brewery, \pi_{\langle \%1, null, null, null \rangle} temp)$

The next rule is based on aggregate constraint $I6$:

> WHEN $INS(beer), DEL(beer)$
> IF NOT $(AVG(beer, alcperc) > 3)$
> THEN $abort$

Finally, the last rule is based on transition constraint $I7$:

> WHEN $INS(beer)$
> IF NOT $(\forall x, y)((x \in beer \land y \in beer_{old} \land x.name = y.name) \Rightarrow$
> $\qquad (x.alcperc \geq y.alcperc))$
> THEN $abort$

This chapter has discussed integrity constraints and integrity rules. These concepts can be seen as the static aspects of integrity control. The next chapter deals with the dynamic aspects of integrity control, i.e. the actual enforcement of integrity constraints to guarantee the integrity of a database.

Chapter 4

Integrity control

The previous two chapters have discussed the general concept of database integrity, and have shown how integrity constraints can be used as a means to explicitly specify the integrity requirements. This chapter describes how integrity constraints can be used in a database management system to actually guarantee the integrity of the database. This task is commonly referred to as *integrity control*.

Integrity control has already received quite some attention in database research. Therefore, this chapter starts with an overview of integrity control as described in database literature, identifying the main subtasks of integrity control and discussing the state of the art in this field. The next section presents the basic ideas of the approach to integrity control described in this thesis; this approach relies heavily on the transaction concept introduced before, and is called *transaction modification*. The ingredients of the transaction modification approach are discussed in the next sections. Finally, the concept of transaction modification is applied in an abstract architecture for a database management system with integrated integrity control. The main components of this architecture coincide with the subtasks of integrity control discussed before.

4.1 An overview of integrity control

This section presents a short overview of integrity control to serve as a background for the sequel of this chapter; the overview is based on the literature survey presented in [Gref91d]. First, history and state of the art of integrity control in relational database management systems are described in a nutshell. Next, the subtasks to be performed by a complete integrity control DBMS subsystem are listed. Finally, an introduction is given to the various approaches to constraint enforcement.

4.1.1 History and state of the art

Below, a short overview of history and state of the art of integrity control in relational database management systems is presented. For further details, the reader is referred to [Vald89, Gref91d]. Note that the discussion below does not include systems that have been fully developed in a commercial context, because details and status of these systems are hard to obtain.

The history of integrity control in relational database management systems started in the mid-seventies with the development of the first full-fledged relational systems. The System R prototype was developed at the IBM San Jose Research Laboratory between 1975 and 1979 [Astr76, Cham81]. An integrated integrity control subsystem for System R is described in [Eswa75]. Although the description of this subsystem is rather complete, the actual implementation of the ideas in System R is rather restricted [Vald89]. Recent versions of the commercial succesor of System R, called DB2, have a more complete functionality [IBM89]. In the System R approach, integrity constraints are specified as predicates in the query language developed in the project, SQL (originally called SEQUEL) [Cham76]. The second important system at the cradle of integrity control is the INGRES system, developed from 1973 to 1980 at the University of California at Berkeley [Stone86a]. This project resulted in the well-known *query modification* method for integrity control [Stone75], to be discussed below. Although the ideas are of a general nature, the actual implementation of the system can handle only rather limited constraint types. Recent commercial versions of INGRES include more general mechanisms [Mark91]. INGRES uses its query language QUEL [Held75] for the specification of integrity constraints.

A more recent relational DBMS incorporating integrity control mechanisms is the SABRE/SABRINA system. The SABRE project started in the early eighties at INRIA and MASI laboratories in France [Gard83]; later, the name of the system was changed to SABRINA. Semantic integrity control has been one of the main research issues in the project [Simon84, Simon87, Gard89]. Integrity constraints are specified as tuple relational calculus assertions or special purpose language constructs for commonly used constraints [Gard89, Vald89]. In [Simon85], the ideas with respect to integrity control are extended to distributed database systems.

The most recent approach to integrity control in relational database management systems is the use of general purpose rule systems for integrity constraint enforcement. The first important project in this context is the successor of the INGRES project, called POSTGRES. POSTGRES is an extended relational DBMS developed during the second half of the eighties at the University of California at Berkeley [Stone86b, Stone88, Stone90a]. Its rule system can be used for several purposes, among which is integrity control. In POSTGRES, integrity constraints are stated in the POSTQUEL language [Rowe87]. The second relevant project is the Starburst project, conducted from the late eighties at IBM Almaden Research Center [Haas90, Lohm91]. Starburst is an extensible DBMS, allowing new functionality to be added to a kernel system. A general purpose rule system is one of the extensions [Widom91]. This rule system can be used for integrity control purposes [Ceri90a, Ceri90b]. In the Starburst approach, integrity constraints are specified in an extension to SQL.

4.1.2 Tasks in integrity control

The integrity control subsystem of a database management system is responsible for all tasks with respect to integrity constraints. The following tasks can be distinguished:

Specification The integrity control subsystem must allow users to specify new constraints and to modify existing constraints.

Analysis Newly defined constraints must be checked for syntactic and semantic correctness before they are accepted by the system.

Completion The integrity control subsystem should complete constraint definitions with the parts that can be generated automatically by the system.

Optimization Constraint definitions are optimized to obtain defintions that are semantically equivalent, but which can be evaluated more efficiently.

Translation Constraint definitions have to be translated from the formalism in which they are specified into the formalism used for constraint enforcement.

Storage Accepted constraints are stored in the system catalog for use at constraint enforcement time. Usually, constraint definitions are stored both in specification formalism and enforcement formalism.

Enforcement When transactions are executed that may violate certain constraints, these constraints have to be enforced.

The emphasis of this thesis is on the constraint enforcement subtask. Therefore, an overview of constraint enforcement mechanisms is presented below.

4.1.3 Constraint enforcement mechanisms

In general, two main approaches to integrity constraint enforcement can be discussed: prevention of incorrect data in the database, or detection and removal of incorrect data from the database. These two approaches are discussed below. The prevention technique is most commonly used; various techniques for the implementation of this technique are therefore discussed.

Violation prevention versus detection

Essentially, there are two basic strategies to enforce integrity constraints [Simon87, Gard89]:

Violation prevention is used if integrity constraints are enforced before the updates of a transaction are actually applied to the database. In case of a violation, the transaction is aborted, and there is no need to undo any changes to the database.

Violation detection is used if integrity constraints are enforced after the updates of a transaction have actually been applied to the database. In case of a violation, the transaction is aborted, and the changes performed by the transaction to the database have to be undone.

From a complexity point of view, the violation detection technique is preferable to the prevention technique, since updates can be applied to the database, the constraints can be evaluated against the database, and standard recovery techniques can be used to undo updates in case of a violation. For this reason, the detection technique is used in most of the earlier systems, for instance DB2 [Vald89]. From a performance point of view, the prevention method is generally preferable, since it eliminates the need to undo changes to a database, which is costly in a disk-based environment. In a main-memory database system, however, this does not necessarily hold, because the database is stored in fast main-memory storage.

Prevention techniques

Various techniques for integrity violation prevention are decsribed in database literature. The most important proposals are described below, in ascending order of general applicability.

Safeness Analysis One approach to constraint violation detection is to analyze update transactions statically to determine whether the transaction is safe. As defined before, a transaction is safe if it can never violate the integrity of a database, regardless of the database state. If the transaction is found to be safe, it can be executed against the database without any further constraint enforcement. If the transaction is found to be unsafe, it is simply rejected and will *never* be executed. In [Gard79], the consistency of transactions is analyzed with Hoare's axiomatic approach to program correctness. In [Wang91] a knowledge-based approach is described. A positive aspect of the technique is that it can be applied at transaction definition time and does not imply any constraint enforcement overhead at transaction execution time. An evident shortcoming of the technique is, however, that it cannot be used for arbitrary transactions and constraints, since it does not take the contents of the database into account. In other words, the set of allowable transactions is severely limited.

Query Modification In the *query modification* approach to constraint enforcement, individual update statements to be executed against the database are modified such that they cannot violate the integrity of the database [Stone75]. Consequently, the modified updates can be executed without any integrity checks. Modification of update statements takes place by extending the update qualification predicate with a predicate derived from relevant constraint definitions. This implies that a modified update statement will in general perform part of the updates specified by the user; this can be regarded as a violation of the transaction atomicity principle.

Transaction Workspace Using the *transaction workspace* approach to constraint violation prevention, the updates to be performed on the database are carried out in a temporary workspace of the transaction. Next, the constraints are evaluated on this workspace and, if necessary, the database. If a constraint violation is detected, the updates in the workspace are simply not propagated to the database. If possible, the transaction workspace is kept in main memory to avoid unnecessary disk access. The transaction workspace method is used in the SABRE/SABRINA project [Simon87, Gard89]. Note that this approach has similarities with the detection approach; they differ in the fact that the workspace approach never allows 'wrong' values in the database itself.

The transaction workspace technique is used in the *transaction modification* approach to integrity constraint enforcement, as discussed in this thesis. The approach is introduced in the next section.

4.2 Transaction modification

This section describes the approach to constraint enforcement of this thesis, called *transaction modification*. The transaction modification approach is based on the following two main ideas:

1. *Enforcement of integrity constraints should as much as possible be integrated with the normal transaction execution mechanisms.*

2. *Enforcement of integrity constraints should take place at the end of transaction execution following the transaction workspace approach.*

These choices have the following advantages:

- Integrity constraint enforcement integrated in normal transaction execution does not require many new mechanisms. From a functional point of view, this ensures simple and clear semantics of integrity control; from a system point of view this ensures simple implementation.

- Transactions including integrity constraint enforcement automatically satisfy the atomicity, serializability and durability requirements of transaction execution.

- Transaction execution mechanisms for performance improvement are automatically available to constraint enforcement; examples are the use of parallelism and advanced scheduling for transaction execution, to be discussed in Chapters 5 and 7, respectively.

Below, an informal introduction is given to the transaction modification technique. Next, an overview of the algorithms necessary for transaction modification is given. These algorithms are treated in detail in the next sections of this chapter.

4.2.1 Introduction to transaction modification

Arbitrary transactions submitted to a database system by a user or application may violate the integrity of the database. The integrity control subsystem of a DBMS should maintain the integrity of the database, regardless of the actions of these transactions. In the transaction modification approach to integrity control, a possible violation of the integrity is prevented by *modifying* each user transaction that contains updates, such that the modified transaction cannot violate the integrity of the database. A transaction is modified by extending it with additional extended relational algebra statements that implement the integrity control task for that transaction. This technique is illustrated below by an example based on the beer database in Appendix B.

Example 4.1 Consider the two integrity rules below, based on referential integrity constraints $I3$, respectively $I4$, of the example database. Both rules specify a cascading delete operation as violation response action [Date83].

$IR1$: WHEN $DEL(brewery)$
IF NOT $(\forall x)(x \in beer \Rightarrow (\exists y)(y \in brewery \land x.brewery = y.name))$
THEN $t1ir1 = unique(\pi_{\langle brewery \rangle} beer) - \pi_{\langle name \rangle} brewery;$
$t2ir1 = beer \bowtie_{\%2=\%5} t1ir1;$
$delete(beer, \pi_{\langle \%1,\%2,\%3,\%4 \rangle} t2ir1)$

$IR2$: WHEN $DEL(beer)$
IF NOT $(\forall x)(x \in trade \Rightarrow (\exists y)(y \in beer \land x.beer = y.name))$
THEN $t1ir2 = unique(\pi_{\langle beer \rangle} trade) - \pi_{\langle name \rangle} beer;$
$t2ir2 = trade \bowtie_{\%2=\%5} t1ir2;$
$delete(trade, \pi_{\langle \%1,\%2,\%3,\%4 \rangle} t2ir2)$

Now suppose a user submits a transaction T_u to the system that removes brewery *Guineken* from the database, and next retrieves the names of all breweries in the

4.2 Transaction modification

database:

$T_u =$ begin
    $delete(brewery, \sigma_{name='Guineken'} brewery)$;
    $?\pi_{\langle name \rangle} brewery$
end

This transaction triggers integrity rule $IR1$ and may violate the condition of this rule. Therefore, transaction T_u is modified with the action of this rule into T_{m1}; the action is appended to the end of T_u.

$T_{m1} =$ begin
    $delete(brewery, \sigma_{name='Guineken'} brewery)$;
    $?\pi_{\langle name \rangle} brewery$;
    $t1ir1 = unique(\pi_{\langle brewery \rangle} beer) - \pi_{\langle name \rangle} brewery$;
    $t2ir1 = beer \bowtie_{\%2=\%5} t1ir1$;
    $delete(beer, \pi_{(\%1,\%2,\%3,\%4)} t2ir1)$
end

The resulting transaction cannot violate rule $IR1$, but the newly added statements may violate rule $IR2$, so a second modification is necessary, resulting in transaction T_{m2}; the action of rule $IR2$ is appended to the end of T_{m1}.

$T_{m2} =$ begin
    $delete(brewery, \sigma_{name='Guineken'} brewery)$;
    $?\pi_{\langle name \rangle} brewery$;
    $t1ir1 = unique(\pi_{\langle brewery \rangle} beer) - \pi_{\langle name \rangle} brewery$;
    $t2ir1 = beer \bowtie_{\%2=\%5} t1ir1$;
    $delete(beer, \pi_{(\%1,\%2,\%3,\%4)} t2ir1)$;
    $t1ir2 = unique(\pi_{\langle beer \rangle} trade) - \pi_{\langle name \rangle} beer$;
    $t2ir2 = trade \bowtie_{\%2=\%5} t1ir2$;
    $delete(trade, \pi_{(\%1,\%2,\%3,\%4)} t2ir2)$
end

This transaction can violate neither of the specified integrity constraints, and can thus be executed without any further integrity control. □

The example above clearly follows the compensating approach to integrity control: 'wrong' values in the database are compensated by additional updates against the database. The emphasis in the sequel of this thesis will, however, be on the aborting approach for two prime reasons. In the first place, the aborting approach is the most general: it can be used for all types of constraints, whereas compensating actions are often hard to design and use. In the second place, the emphasis on the aborting approach is of a pragmatic nature, because it reduces the number of techniques to be discussed in the sequel of this thesis. Integrity control through transaction modification as presented in this and the following chapters is perfectly fit for both the compensating and the aborting approach, however.

4.2.2 Transaction modification subtasks

From the informal description of transaction modification given above and the general list of tasks for integrity control presented in the first section of this chapter, the main subtasks of an integrity control subsystem using the transaction modification technique can be deduced.

Integrity specification In the first place, the subsystem must allow users to specify new integrity constraints or rules, or modify existing ones. Further, the system should check the validity of the rules and fill in missing parts. This subtask is discussed in detail in Section 4.3.

Integrity rule preprocessing As integrity requirements are specified as rules, but are enforced by means of extended relational algebra constructs, a translation of rules into extended relational algebra is necessary. Further, rules should be optimized to reduce the cost of integrity control. Rule preprocessing is dealt with in Section 4.4.

Constraint enforcement Update transactions submitted to the system must be modified as illustrated before. This requires algorithms to select the rules that may be violated by the transaction and to actually modify the transaction with preprocessed forms of these rules. Further, the modified transaction must be executed. Constraint enforcement is treated in Section 4.5.

4.3 Integrity specification

This section deals with integrity specification, i.e. the mechanisms necessary for the definition of new integrity constraints or rules and the modification of existing constraints or rules. If the integrity control subsystem requires the explicit specification of integrity violation response actions, integrity rules must be defined; otherwise, integrity constraints must be defined. The specification mechanisms include three aspects:

Interface In the first place, the integrity control subsystem must offer an interactive interface for the actual specification of integrity constraints or rules. In the case of rules, conditions and actions have to be specified, and optionally trigger sets.

Validation Secondly, the integrity control subsystem must validate new constraints or rules, i.e. ensure that they are syntactically and semantically correct.

Completion Finally, the subsystem must complete integrity rules with the parts not specified by the user. Completion of rules should at least support the

4.3 Integrity specification

generation of trigger sets. In some cases, the generation of default violation response actions may be possible for some constraint types[1].

These three aspects of integrity specification are discussed in detail below.

4.3.1 Integrity specification interface

An integrity control subsystem can offer various user interfaces for the interactive specification of integrity constraints or rules. Several classes of interfaces can be distinguished:

Language-oriented In the language-oriented approach, constraints are specified in some kind of data definition language. This can either be an extension to the general data definition language of the system or a special-purpose constraint definition language.

Form-oriented In the form-oriented approach, constraints are specified by filling in conditions in forms generated by the system. Mostly, these forms are a representation of existing relations, like in the QBE system [Zloof75, Zloof78].

Toolbox-oriented The toolbox-oriented approach allows the construction of constraint specification with the use of toolboxes providing building blocks and construction primitives for constraints. An example of this approach can be found in the SuperBase system [Super90].

Graphics-oriented In the graphical approach, constraints are specified by arranging graphical symbols representing constraint constructs in a graphical workspace. An example in the context of entity-relationship schema definition is the graphical schema editor of the SUPER database visual environment [Auddi91].

The language-oriented approach is the most simple and general. Given a language-oriented interface, another type of interface can be used on top of this interface to offer a user-friendly interface for all or certain classes of constraints or rules. Here, we will assume that the integrity specification process results in rule specifications in the $\mathcal{RL}$ language, where the trigger sets are to be filled later by the system.

4.3.2 Integrity rule validation

As discussed above, integrity rules are specified in the $\mathcal{RL}$ language. Before the rules can be accepted by the system, they have to be verified. The verification of a new integrity rule J includes the following issues [Tsich82, Brod78]:

---

[1] Examples are domain constraints, where 'illegal' values can be updated to null values by default, and referential integrity constraints with cascading update semantics [Date83].

- Checking the syntactic correctness or *well-formedness* of J with respect to the $\mathcal{RL}$ language. This issue may in fact be considered part of integrity rule specification instead of verification.

- Checking the semantic correctness of the individual rule J. Relevant aspects are for example: do all relations and attributes used in J exist, and are comparisons made only between compatible operands[2]?

- In the case that the condition of J is a state constraint, checking if this constraint is not violated by the database state D at the time J is defined. This requires evaluating the following condition:

$$condition(J)(D)$$

If the constraint does not hold at definition time, it is likely that no update on the database will succeed anymore.

- Checking if the condition of J is not implied by the set of conditions of already existing rules $\mathcal{J}$, i.e. checking the minimality of the constraint set:

$$\neg\left(\bigwedge_{J' \in \mathcal{J}} condition(J') \Rightarrow condition(J)\right)$$

If the condition evaluates to false, the new constraint is superfluous. Although a redundant set of constraints may be semantically correct, it is undesirable from viewpoints of efficiency and constraint maintenance.

- Checking if the condition of J is not contradicted by the set of conditions of already existing rules $\mathcal{J}$, i.e. checking the consistency of the constraint set [Bry86]:

$$\neg\left(\bigwedge_{J' \in \mathcal{J}} condition(J') \Rightarrow \neg condition(J)\right)$$

If the set of constraints is inconsistent, no valid database states or transitions can exist if the constraint is accepted. Note that for a state constraint, checking the constraint succesfully against the current database state is sufficient to prove that there is no contradiction.

Integrity control subsystems usually implement part of the verification described above. The first three verification steps are essential in a real-world system. The last two steps are desirable, but hard to realize.

---

[2] This typing issue is related to the concept of *underlying domain* as introduced in the ALPHA language [Codd71]. This concept allows the database designer to specify that certain attributes are not compatible, even though they have the same basic domain.

4.3 Integrity specification

4.3.3 Trigger set generation

As described above, the user specifies integrity rule conditions and optionally integrity violation response actions. The system is responsible for the generation of trigger sets to obtain complete integrity rules as defined in Chapter 3. A trigger set for an integrity rule is derived from the condition of the rule. In this section we describe the generation of trigger sets for constraints specified in the $\mathcal{CL}$ constraint specification language. A similar generation of triggers for constraints specified in an SQL-like syntax is described in [Ceri90b].

Algorithm 4.1 A trigger set is generated from an integrity rule condition in the form of a $\mathcal{CL}$ well-formed formula W by function $GenTrigC$ as shown below. In this definition, the symbols V_u and V_e denote the set of universally respectively existentially quantified variables in a condition.

$$GenTrigC(W) = GenTrigW(W, \emptyset, \emptyset)$$

$GenTrigW(W, V_u, V_e) =$

$$\begin{cases} GenTrigW(W_1, V_u \cup \{x\}, V_e) & \text{if } W = (\forall x)W_1 \\ GenTrigW(W_1, V_u, V_e \cup \{x\}) & \text{if } W = (\exists x)W_1 \\ GenTrigW(W_1, V_u, V_e) \cup GenTrigW(W_2, V_u, V_e) & \text{if } W = W_1 \phi W_2 \\ GenTrigN(W_1, V_u, V_e) \cup GenTrigW(W_2, V_u, V_e) & \text{if } W = W_1 \Rightarrow W_2 \\ GenTrigN(W_1, V_u, V_e) & \text{if } W = \neg W_1 \\ GenTrigA(W, V_u, V_e) & \text{otherwise} \end{cases}$$

$GenTrigN(W, V_u, V_e) =$

$$\begin{cases} GenTrigN(W_1, V_u, V_e \cup \{x\}) & \text{if } W = (\forall x)W_1 \\ GenTrigN(W_1, V_u \cup \{x\}, V_e) & \text{if } W = (\exists x)W_1 \\ GenTrigN(W_1, V_u, V_e) \cup GenTrigN(W_2, V_u, V_e) & \text{if } W = W_1 \phi W_2 \\ GenTrigW(W_1, V_u, V_e) \cup GenTrigN(W_2, V_u, V_e) & \text{if } W = W_1 \Rightarrow W_2 \\ GenTrigW(W_1, V_u, V_e) & \text{if } W = \neg W_1 \\ GenTrigA(W, V_u, V_e) & \text{otherwise} \end{cases}$$

$GenTrigA(A, V_u, V_e) =$

$$\begin{cases} GenTrigT(T_1) \cup GenTrigT(T_2) & \text{if } A = T_1 \vartheta T_2 \\ \{\langle INS, R \rangle\} & \text{if } A = (x \in R) \text{ and } x \in V_u \\ \{\langle DEL, R \rangle\} & \text{if } A = (x \in R) \text{ and } x \in V_e \\ \emptyset & \text{otherwise} \end{cases}$$

$(\forall x)(x \in R)$	$t \leftarrow t \cup \{\langle INS, R\rangle\}$
$(\exists x)(x \in R)$	$t \leftarrow t \cup \{\langle DEL, R\rangle\}$
$aggr(R, i)$	$t \leftarrow t \cup \{\langle INS, R\rangle, \langle DEL, R\rangle\}$
$cnt(R)$	$t \leftarrow t \cup \{\langle INS, R\rangle, \langle DEL, R\rangle\}$

Table 4.1: Trigger set generation from constraint language (informal)

$GenTrigT(T) =$

$$\begin{cases} \{\langle INS, R\rangle, \langle DEL, R\rangle\} & \text{if } T = \Gamma_1(R, i) \\ \{\langle INS, R\rangle, \langle DEL, R\rangle\} & \text{if } T = \Gamma_2(R) \\ \emptyset & \text{otherwise} \end{cases}$$

where $\phi \in \{\wedge, \vee\}, \vartheta \in \{<, \leq, =, \neq, \geq, >\},$
$\Gamma_1 \in \{SUM, AVG, MIN, MAX\}, \Gamma_2 \in \{CNT, MLT\}$

$\triangledown$

Function $GenTrigC$ breaks down a condition following the $\mathcal{CL}$ syntax presented in Chapter 3. The relation between $\mathcal{CL}$ constructs and a trigger set t is summarized informally in Table 4.1.

With respect to triggers generated from aggregate functions, some optimizations are possible by detailed inspection of the aggregate function and the predicate it is used in. This requires a more semantics oriented analysis of the constraint definitions, which is more complicated than the simple syntactic analysis as presented above. This kind of optimization is discussed extensively in [Ceri90b].

4.4 Integrity rule optimization and translation

After new or modified integrity rules have been accepted by the system as discussed in the previous section, they are preprocessed for use in constraint enforcement. Preprocessing of rules consists of two parts:

- Optimization of integrity rules: conditions and actions of the integrity rules are optimized to obtain equivalent rules that can be enforced at lower cost.

- Translation of rules: integrity rules are translated to extended relational algebra programs. The translated form can directly be used for constraint enforcement as discussed in Section 4.5.

Optimization and translation functions for rules are discussed below. These algorithms are then used for the definition of a function that preprocesses a set of integrity rules to be used for transaction modification.

4.4 Integrity rule optimization and translation

4.4.1 Integrity rule optimization

This section discusses the optimization of integrity rules, i.e. the transformation of a rule J into a rule J' that has the same semantics, but can be evaluated at a lower cost. Complete optimization of a rule consists of optimization of the condition and the action of the rule. Here we focus on optimization of the condition for two reasons. In the first place, optimization of relational algebra constructs is dealt with extensively in the field of query optimization; techniques developed in this context can be used for the optimization of integrity rule actions. In the second place, as mentioned before, the emphasis of this thesis is on the aborting approach to integrity control; in this approach there is no action that can be optimized.

Algorithm 4.2 Restricting the optimization of integrity rules to the optimization of the condition of the rule, the optimization function $OptR$ can be defined as follows, where J is an integrity rule:

$$OptR(J) = \langle triggers(J), OptC(condition(J)), action(J) \rangle$$

Here, function $OptC$ transforms a condition into an equivalent optimized condition. ▽

Function $OptC$ will not be further specified in full detail. The functionality of the function can be chosen freely within the boundaries of the equivalence criterium, depending on the required level of optimization. Specification of a 'complete' function is an important research area. Several techniques to be used in $OptC$ are described below informally.

Differential relations

Given the fact that a pre-transaction state of a database is correct, a reduction of the amount of data to be checked in constraint enforcement can be obtained by inspecting only those parts of relations that have been changed in a relevant way by a transaction. This is usually accomplished by the use of *differential relations* [Simon87, Gard89, Gref89a, Gref90a]. In this approach, two auxiliary relations are associated with each base relation R:

new differential relation The new differential relation contains tuples that have been inserted and the new values of tuples that have been modified by the current transaction. The new differential relation associated with relation R is usually denoted as R^+.

old differential relation The old differential relation contains tuples that have been deleted and the old values of tuples that have been modified by the current transaction. The old differential relation is usually denoted as R^-.

Using these auxiliary relations, constraint conditions can be reformulated.

Example 4.2 An example is domain constraint $I1$ from the example database:

$$(\forall x)(x \in beer \Rightarrow x.alcperc \geq 0)$$

Given the fact that the constraint holds on the pre-transaction state of relation *beer*, only the new tuples in *beer* have to be checked. Therefore, the constraint can be optimized as follows:

$$(\forall x)(x \in beer^+ \Rightarrow x.alcperc \geq 0)$$

□

Syntactic manipulation

In [Nico82] the simplification of integrity constraints in first-order logic is described. The basic principle of the described method relies on the instantiation of formulas obtained by substituting for some of their variables the (or some of the) constants occurring in the inserted (or deleted, or updated) tuple. The technique is only applicable to transactions updating multiple tuples if the order of the operations in a transaction is immaterial.

In [Hsu85] the simplification of constraints in the context of general transactions is described. The simplification is based on the prefix of the constraint (quantifiers).

Semantic manipulation

A knowledge-based approach to constraint optimization and enforcement is described in [Qian86]. In this approach, a constraint manager keeps a knowledge base with semantic knowledge about application and database. The knowledge is extracted from the database schema definition, from user-defined constraints, and from the results of continuous monitoring of the database state. The knowledge is used to optimize and enforce constraints.

4.4.2 Integrity rule translation

This section deals with the translation of integrity rules into extended relational algebra programs. In general, a program is derived from the condition and action of an integrity rule. We can distinguish between rules with an aborting character and rules with a compensating character. If the rule has an aborting character, only the condition of the rule has to be translated to extended relational algebra constructs. So, we have the following.

4.4 Integrity rule optimization and translation

Algorithm 4.3 Function *TransR* translates an integrity rule J into an extended relational algebra program:

$$TransR(J) = \begin{cases} TransC(condition(J)) & \text{if } action(J) = abort \\ TransCA(condition(J), action(J)) & \text{otherwise} \end{cases}$$

$\triangledown$

With respect to function *TransCA* some remarks can be made. In most practical cases, the program produced by this function is equal to the violation response action given as argument to the function. Other cases require a deeper analysis of condition and action; this analysis is beyond the scope of this thesis. Consequently, this section deals with function *TransC*, i.e. the translation of conditions in the $\mathcal{CL}$ language into the extended relational algebra.

The extended relational algebra defined in Chapter 2 includes the constructs necessary for the construction of compensating programs. For the construction of aborting programs, however, a new construct has to be added. This construct is described and defined below. Next, the translation of constraints is discussed.

Alarm operator

In database systems, an abort operator is used to undo all effects of a transaction in case of an error. Normally, this abort operator is used by the system in situations like concurrency control deadlock or media failure; it can also be used by the user to undo the effects of a transaction in execution.

In the transaction modification approach to integrity control, the abort operator is used explicitly to undo transactions the completion of which would imply an integrity violation. This requires a new operator that raises an abort state if some condition on the database holds. A simple operator is chosen here, that can easily be integrated into the extended relational algebra. The operator, *alarm*, accepts an arbitrary relational expression as its operand, and generates an abort situation if this operand is nonempty. The following definition adds the *alarm* operator to the extended relational algebra by redefining the statements defined in Chapter 2.

Definition 4.1 The *extended relational algebra statements* are redefined as follows. Let E denote an arbitrary extended relational expression as defined in Definition 2.9. Then the following constructs are extended relational algebra statements:

- The extended relational algebra statements as defined in Definition 2.10.

- The *alarm* statement $alarm(E)$ aborts the transaction it belongs to if E is non-empty; otherwise it has no effect:

$$alarm(E) \equiv \begin{cases} abort & \text{if } CNT(E,1) > 0 \\ nothing & \text{otherwise} \end{cases}$$

Constraint translation

As described above, constraints specified in the $\mathcal{CL}$ language have to be translated to the extended relational algebra for enforcement by function *TransC*. A complete specification of a translation algorithm is not given here; research remains to be performed into the translation of constructs including multi-sets and aggregates. A starting point for a general approach and the translation of a number of important cases are presented below.

Constraints consisting of a quantified formula form an important class of constraints; for this class, function *TransC* can be defined as follows.

Algorithm 4.4 Let c denote a constraint (well-formed formula) in the $\mathcal{CL}$ language, and assume that c has the form $(\vartheta x)(c'(x))$, with $\vartheta \in \{\forall, \exists\}$. Then function *TransC* translates c into an aborting extended relational algebra construct:

$TransC(c) =$

$$\begin{cases} alarm(CalcToAlg(\{x \mid \neg c'(x)\})) & \text{if } c = (\forall x)(c'(x)) \\ alarm(\sigma_{\%1=0}(groupby(\langle\rangle, CNT, \%1, \\ \qquad CalcToAlg(\{x \mid c'(x)\})))) & \text{if } c = (\exists x)(c'(x)) \end{cases}$$

$\triangledown$

Constraints consisting of a negated quantified formula can easily be converted to a non-negated form. Simple aggregate constraints form a separate class, the translation of which is easy (see below). Function *CalcToAlg* translates a $\mathcal{CL}$-based relational tuple calculus expression into the equivalent extended relational algebra expression. A complete algorithm for this function is not given here, but some important remarks with respect to it are discussed.

To construct an algorithm for function *CalcToAlg*, we must be able to map every construct of $\mathcal{CL}$ to the extended relational algebra. A formal proof of this equivalence between the languages is not given here; a line of reasoning is presented below that makes the equivalence acceptable:

- The $\mathcal{CL}$ language is based on the standard tuple relational calculus.
- The extended relational algebra is based on the standard relational algebra.
- The equivalence between standard tuple relational calculus and standard relational algebra is shown in [Ullm82].
- The extensions of $\mathcal{CL}$ with respect to the standard tuple relational calculus are multi-set semantics and aggregate functions; these extensions are also included in the extended relational algebra.

Translation algorithms from the standard tuple calculus to the standard relational algebra are discussed in the literature, see for example [Pare89]. These algorithms can be used as a starting point for the construction of function *TransC*.

4.4 Integrity rule optimization and translation

C	$(\forall x)(x \in R \Rightarrow c(x))$
$TransC(c)$	$alarm(\sigma_{\neg c'} R)$
	$(\forall x)(x \in R \Rightarrow (\exists y)(y \in S \wedge x.i = y.j))$
	$alarm(unique(\pi_{\langle \%i \rangle} R) - \pi_{\langle \%j \rangle} S)$
	$(\forall x)(x \in R \Rightarrow (\forall y)(y \in S \Rightarrow x.i \neq y.j))$
	$alarm(\pi_{\langle \%i \rangle} R \cap \pi_{\langle \%j \rangle} S)$
	$(\forall x,y)((x \in R \wedge y \in S \wedge c_1(x,y)) \Rightarrow c_2(x,y))$
	$alarm(\sigma_{\neg c'_2}(R \bowtie_{c'_1} S))$
	$(\exists x)(x \in R \wedge c(x))$
	$alarm(\sigma_{\%1=0}(groupby(\langle\rangle, CNT, \%1, \sigma_{c'} R)))$
	$c(AGGR(R,i))$
	$alarm(\sigma_{\neg c'}(groupby(\langle\rangle, AGGR, \%i, R)))$
	$c(CNT(R))$
	$alarm(\sigma_{\neg c'}(groupby(\langle\rangle, CNT, \%1, R)))$

Table 4.2: Translation of typical constraint constructs

The translation of a number of typical constructs in $\mathcal{CL}$ is shown in Table 4.2. In this table, c, c_1, and c_2 denote conditions defined on a tuple in $\mathcal{CL}$ format. The corresponding conditions c', c'_1, and c'_2 in the extended relational algebra are obtained by trivial syntactical modification. Further, $AGGR$ denotes an arbitrary aggregate function from $\mathcal{CL}$, and $COUNT$ a counting function.

Translation of example constraints

The translation of the constraints of the example beer database (see Appendix B) is presented in Table 4.3. The examples show that constraints can easily be translated to simple extended relational algebra constructs. Both state and dynamic constraints can be handled. Note that the translation of a tuple-oriented transition constraint like $I7$ requires a join because of the set-oriented character of the algebra.

Alternative translations of constraints are possible, and the translation of a single constraint may consist of multiple extended relational algebra statements, as demonstrated by the following example.

Example 4.3 Uniqueness constraint $I2$ from the example database is defined as follows:

$(MLT(brewery \leq 1)) \wedge$
$(\forall x, y)((x, y \in brewery \wedge x.name = y.name) \Rightarrow (x = y))$

Table 4.3 shows a translation of this constraint that checks whether the attribute *name* of relation *brewery* contains the same values before and after duplicate

I	$TransC(I)$
$I1$	$alarm(\sigma_{\neg alcperc \geq 0} beer)$
$I2$	$alarm(\pi_{\langle name \rangle} brewery - unique(\pi_{\langle name \rangle} brewery))$
$I3$	$alarm(unique(\pi_{\langle brewery \rangle} beer) - \pi_{\langle name \rangle} brewery)$
$I4$	$alarm(unique(\pi_{\langle beer \rangle} trade) - \pi_{\langle name \rangle} beer)$
$I5$	$alarm(\sigma_{\neg qty\_bought \geq qty\_sold} trade)$
$I6$	$alarm(\sigma_{\neg \%1 > 3} gb(\langle \rangle, AVG, alcperc, beer))$
$I7$	$alarm(\sigma_{\neg \%4 \geq \%8}(beer \bowtie_{\%1=\%5 \wedge \%2=\%6} beer\_old))$

Table 4.3: Translation of example constraints

removal. An alternative multi-statement translation is the following:

$t_1 = CNT(brewery);$
$t_2 = CNT(unique(\pi_{\langle name \rangle} brewery));$
$alarm(\sigma_{\%1 \neq \%2}(t_1 \times t_2))$

$\square$

4.4.3 Rule set preprocessing

Above, functions have been described for the optimization and translation of individual integrity rules. For the definition of a transaction modification function, we must be able to handle multiple rules that may be violated by a given transaction. For this reason a rule set preprocessing function is defined.

Algorithm 4.5 Given a set of integrity rules $\mathcal{J}$, function $TrOptRS$ preprocesses these rules to produce a single extended relational algebra program:

$$TrOptRS(\mathcal{J}) = \begin{cases} P_\varepsilon & \text{if } \mathcal{J} = \emptyset \\ TransR(OptR(head(\mathcal{J}))) \oplus TrOptRS(tail(\mathcal{J})) & \text{otherwise} \end{cases}$$

$\triangledown$

Note that this algorithm interprets the set of integrity rules as a list to enable the use of the *head* and *tail* operators; the set can be converted to a list by defining an arbitrary order on its elements. Function $TrOptRS$ translates and optimizes a rule set (this accounts for the name of the function) by concatenating the extended relational algebra programs resulting from optimization and translation of the individual rules.

4.5 Constraint enforcement

This section deals with the enforcement of integrity constraints. This task consists of three subtasks:

- Selection of the integrity rules of which the conditions may be violated by a given user transaction; the selection is based on the trigger sets of the rules.

- Modification of the user transaction according to the transaction modification technique introduced before.

- Actual enforcement of the constraints through the execution of the transaction modifier that is composed of translated forms of the selected constraints.

4.5.1 Integrity rule selection

Constraint selection requires two mechanisms: a mechanism to derive the triggers from an extended relational algebra program, and a mechanism to match these triggers with the triggers of a set of integrity rules. Functions for these two mechanisms are descibed below.

Trigger set derivation

The derivation of a set of triggers from an extended relational algebra program is performed by a simple syntactical analysis of the program. The function *GetTrigP* shown below describes the basic functionality of this analysis. The function produces a set of triggers, i.e. the derived trigger set does not contain duplicate triggers.

Algorithm 4.6 The trigger set derivation function *GetTrigP* derives a set of triggers from an extended relational algebra program; it is defined as follows:

$$GetTrigP(P) = \begin{cases} \emptyset & \text{if } P = P_\varepsilon \\ GetTrigS(head(P)) \cup GetTrigP(tail(P)) & \text{otherwise} \end{cases}$$

where

$$GetTrigS(S) = \begin{cases} \{\langle INS, R \rangle\} & \text{if } S = insert(R, E) \\ \{\langle DEL, R \rangle\} & \text{if } S = delete(R, E) \\ \{\langle INS, R \rangle, \langle DEL, R \rangle\} & \text{if } S = update(R, E, \alpha) \\ \emptyset & \text{otherwise} \end{cases}$$

$\triangledown$

Function *GetTrigP* recursively traverses the statements of an extended relational algebra program, and adds the appropriate trigger to the resulting trigger set whenever it encounters a statement that modifies a relation.

Integrity rule selection

In the transaction modification approach, integrity rules are selected by deriving a set of triggers from an extended relational algebra program and matching this set with the trigger sets of the integrity rules defined on the database. Using function *GetTrigP* defined above, we can now construct the integrity rule selection function *SelRS*.

Algorithm 4.7 The integrity rule selection is performed by function *SelRS*; this function selects a set of integrity rules from a set $\mathcal{J}$ based on the statements in extended relational algebra program P:

$$SelRS(P, \mathcal{J}) = \{J \in \mathcal{J} \mid triggers(J) \cap GetTrigP(P) \neq \emptyset\}$$

$\triangledown$

4.5.2 Transaction modification

Having defined the functions for integrity rule optimization, translation, and selection, we can proceed to the top level algorithms of the transaction modification technique. To modify transactions, a few simple operators are necessary to manipulate transactions; these operators are discussed first. Next, the actual transaction modification function is described in detail; this function forms the kernel of the transaction modification technique. Finally, some attention is paid to the actual enforcement of constraints, i.e. to the execution of modified transactions.

Transaction and program operators

When constraint definitions stated as an extended relational algebra program have to be added to a transaction for enforcement, operators are necessary to manipulate transactions and programs. These operators are introduced below.

Definition 4.2 The *transaction debracketing operator* transforms a transaction into an extended relational algebra program by removing the transaction brackets. Let T denote a transaction, then $T\downarrow$ is defined as follows:

$T = (a_1; \ldots; a_n)$
$T\downarrow = a_1; \ldots; a_n$

Here, the brackets denote the transaction brackets *begin* and *end*. □

Definition 4.3 The *program bracketing operator* transforms an extended relational algebra program into a transaction by adding the transaction brackets. Let P denote an extended relational algebra program, then $P\uparrow$ is defined as follows:

$P = a_1; \ldots; a_n$
$P\uparrow = (a_1; \ldots; a_n)$

Again, the brackets denote the transaction brackets *begin* and *end*. □

4.5 Constraint enforcement

Transaction modification function

Now all ingredients are available for the definition of the transaction modification function *ModT*. This function modifies an arbitrary user transaction such that the execution of this transaction will not violate a given set of integrity rules.

Algorithm 4.8 Let T denote an arbitrary transaction and $\mathcal{J}$ a set of integrity rules defined on a database. Then the *modified transaction* of T with respect to $\mathcal{J}$ is defined recursively by the function *ModT* as follows:

$ModT(T, \mathcal{J}) = ModP(T\downarrow, \mathcal{J})\uparrow$
where
$ModP(P, \mathcal{J}) = \begin{cases} P & \text{if } TrigP(P, \mathcal{J}) = P_\varepsilon \\ P \oplus ModP(\,TrigP(P, \mathcal{J}), \mathcal{J}) & \text{otherwise} \end{cases}$
and
$TrigP(P, \mathcal{J}) = TrOptRS(SelRS(P, \mathcal{J}))$

$\triangledown$

As discussed before, function *TrOptRS* translates a set of rules into an extended relational algebra program; function *SelRS* selects a subset from a set of integrity rules based on the statements in an extended relational algebra program, and returns the selected rules in a list. Recall that P_ε denotes the empty extended relational algebra program. Termination of the transaction modification algorithm depends on the defined integrity rules; this issue will be discussed later (Section 4.6.1).

The extended relational algebra program added to a transaction for integrity control purposes is called the transaction modifier.

Definition 4.4 Let T denote an arbitrary transaction and $\mathcal{J}$ a set of integrity rules. Then the *transaction modifier* of T with respect to $\mathcal{J}$ is defined as:

$ModP(\,TrigP(T\downarrow, \mathcal{J}), \mathcal{J})$

A transaction modifier can be considered layered, where each invocation of *ModP* adds one layer to the transaction modifier; the program of the unmodified transaction is called layer 0. For a transaction T the layers are thus defined as follows:

$layer0$: $T\downarrow$
$layer1$: $TrigP(T\downarrow, \mathcal{J})$
$layer2$: $TrigP(\,TrigP(T\downarrow, \mathcal{J}), \mathcal{J})$
$\vdots$ $\vdots$

$\square$

Depending on the violation response action in rule set $\mathcal{J}$, the transaction modifier implements an aborting, compensating, or mixed approach to integrity control: if the violation response actions contain no update, insert, or delete statements, the approach is purely aborting; if the actions contain no alarm statements, the approach is purely compensating. As stated before, the emphasis of this thesis is on the aborting approach. In this case, the function *ModP* defined above degenerates to the following:

$$ModP(P, \mathcal{J}) = P \oplus TrigP(P, \mathcal{J})$$

Comparison to other approaches

The transaction modification technique is based on analysis of a transaction specification, and not on analysis of the net effects of transaction execution. As such, transaction modification is a pessimistic approach to integrity control, because actions are taken against *possible* database modifications, not against *actual* modifications. This implies that constraint enforcement will be activated more often than strictly necessary, which can be considered an inefficiency of the technique. The triggering algorithms can even lead to infinite triggering behaviour; this problem is discussed in Section 4.6.1. Other approaches, like the rule-based technique used in the Starburst context [Ceri90b, Lohm91], are based on the net effect of transaction execution [Widom90, Widom91]. These approaches trigger constraint enforcement upon actual database changes. On the other hand, the transaction modification technique causes little overhead and has a number of other performance-related advantages that will be discussed in the sequel of this thesis.

An approach that has some similarities with transaction modification is the *query rewrite* technique, used in the second version of the POSTGRES rule system [Stone90a, Stone90b]. In this approach, a user command is rewritten by integrating it with the actions of rules triggered by the command; the technique can be considered an extension of the *query modification* technique used in the INGRES system [Stone75]. There are a number of important differences between the transaction modification and query rewrite techniques, however. In the first place, the query rewrite technique operates per user statement, i.e. transactions are not taken into account at all. In the second place, the query rewrite technique can only be used for rules with single-statement actions[3] [Stone90b], whereas transaction modification allows arbitrary actions. Finally, the query rewrite technique requires a more complex analysis of user commands and rules at query rewrite time than in the transaction modification case at modification time.

4.5.3 Constraint execution

The transaction modification technique translates integrity constraints to relational algebra constructs that can be fully integrated into a normal transaction. Consequently, no special mechanisms for constraint execution are necessary.

---

[3]The POSTGRES system is equipped with a second, tuple-based implementation of the rule system, that can handle arbitrary actions.

4.6 Operational aspects

Other approaches require mechanisms that can handle rule execution (see e.g. [Stone88, Widom91]).

As constraints are enforced fully within the transaction execution mechanism, the transaction properties guaranteed by the system are automatically valid for constraint execution as well. This means that:

- the execution of a modified transaction is atomic;
- the execution of a modified transaction is serializable;
- the durability of the effects of a modified transaction is guaranteed.

The clear and simple semantics of the transaction modification technique allow for much freedom in the actual implementation of the technique in a database management system. From a semantical point of view there is no difference between the situation in which a transaction is first completely modified and then executed, and the situation in which each layer of a modified transaction is generated and executed separately. From an implementation point of view, each situation has certain advantages. The first situation allows for stored pre-modified transactions, whereas the second situation allows for parallel transaction modification and execution, for example.

4.6 Operational aspects

The sections above have discussed the concepts behind the transaction modification technique. Some operational aspects require some more attention, however, to make the ideas more usable in practice. The most important aspects are discussed below.

4.6.1 Infinite triggering suppression

The fact that the transaction modification technique uses a static analysis to select the triggered integrity rules creates a relatively high risk of infinite triggering behaviour. This is illustrated by the following example.

Example 4.4 The following integrity rule can be associated with domain constraint $I1$ from the example database:

WHEN $INS(beer)$
IF NOT $(\forall x)(x \in beer \Rightarrow x.alcperc \geq 0)$
THEN $update(beer, \sigma_{alcperc<0} beer, \langle name, brewery, type, 0 \rangle)$

This rule sets all illegal values in the $alcperc$ attribute of relation $beer$ to a default value. This update will, however, always trigger the rule again, regardless of whether tuples have actually been updated. Consequently, the rule will trigger itself infinitely. □

The triggering behaviour of a set of integrity rules can be analyzed by means of a triggering graph as defined below.

Definition 4.5 Let $\mathcal{J}$ denote a set of integrity rules. Than the *triggering graph* of $\mathcal{J}$ is a directed graph $G = \langle V, E \rangle$. The set V denotes the vertices of G and corresponds with the set of integrity rules $\mathcal{J}$. The set E denotes the edges of G and is defined as follows:

$$E = \{\langle J_1, J_2 \rangle \mid J_1, J_2 \in V \wedge GetTrigP(action(J_1)) \cap triggers(J_2) \neq \emptyset\}$$

□

Infinite rule triggering in a rule set $\mathcal{J}$ can only occur if the triggering graph of $\mathcal{J}$ contains one or more cycles. A simple way to remove cycles is to allow actions of integrity rules to be declared as *non-triggering* extended relational algebra programs.

Definition 4.6 A *non-triggering relational algebra program* P_s will never trigger any rule. The trigger set derivation function $GetTrigP$ as discussed before is replaced by function $GetTrigPX$:

$$GetTrigPX(P) = \begin{cases} \emptyset & \text{if } non\text{-}triggering(P) \\ GetTrigP(P) & \text{otherwise} \end{cases}$$

□

Given these concepts, an integrity control subsystem can validate a set of integrity rules with respect to triggering behaviour by constructing and analyzing the triggering graph. If cycles are detected, the system assists the user in removing these cycles. This approach is comparable to that described in [Ceri90a, Ceri90b]. It does, however, place a heavy burden on the shoulders of the database designer, since the system cannot guarantee integrity completely automatically in this approach.

The approach described above can be used to avoid 'unnecessary' infinite triggering behaviour. Of course it is possible to define a set of integrity rules that inherently imply an infinite triggering process. Clearly, such a set of rules has to be considered semantically incorrect.

4.6.2 Static optimization and translation

As specified by the definition of the transaction modification function $ModT$, integrity rules are optimized and translated each time a transaction is modified. Clearly, this is not necessary, as rules can be optimized and translated once when they are specified. The translated form is then stored for use at constraint enforcement time. This also requires the trigger set of a rule to be stored with a translated rule. Therefore, the concept of integrity program is defined below.

4.6 Operational aspects

Definition 4.7 An *integrity program* defined on a database schema $\mathcal{D}$ is a pair $K = \langle t, p \rangle$ with the following elements:

- The *trigger set* t is a set of pairs $\langle u, r \rangle$ with $u \in \{INS, DEL\}$ and r a relation name in $\mathcal{D}$.
- The *triggered program* p is an extended relational algebra program.

If $K = \langle t, p \rangle$ denotes an integrity program, $triggers(K)$ denotes the trigger set t, and $action(K)$ denotes the triggered program p. □

This definition of an integrity program can easily be extended with a flag indicating whether the program is non-triggering as discussed above.

As may be clear by now, an integrity program is derived from an integrity rule J using the rule optimization and translation functions presented before.

Algorithm 4.9 The integrity program generation function $GetIntP$ is defined as follows, where J is an integrity rule:

$$GetIntP(J) = \langle triggers(J), TransR(OptR(J)) \rangle$$

▽

Given the fact that integrity rules are translated at rule definition time and stored in a set of integrity programs, the transaction modification algorithm has to be adapted. The changes to the algorithm are rather straightforward:

Algorithm 4.10 Function $TrigP$ from the transaction modification algorithm is redefined as follows. Here P is an extended relational algebra program and $\mathcal{K}$ a set of integrity programs.

$$TrigP(P, \mathcal{K}) = ConcatP(SelPS(P, \mathcal{K}))$$

where

$$SelPS(P, \mathcal{K}) = \{K \in \mathcal{K} \mid triggers(K) \cap GetTrigP(P) \neq \emptyset\}$$

$$ConcatP(\mathcal{K}) = \begin{cases} P_\varepsilon & \text{if } \mathcal{K} = \langle\rangle \\ action(head(\mathcal{K})) \oplus ConcatP(tail(\mathcal{K})) & \text{otherwise} \end{cases}$$

▽

As may be clear, function $SelPS$ is the substitute for function $SelRS$, and $ConCatP$ is the substitute for $TrOptRS$. Note that rule optimization and translation are not included in $ConCatP$, since these tasks are not performed at constraint enforcement time. Note further that the definition above interprets the set $\mathcal{K}$ as a list;

as mentioned bofore, a set can be interpreted as a list by imposing an arbitrary order on the elements of the set.

4.7 Abstract system architecture

In this section an abstract system architecture is presented that can be used for handling integrity constraints through transaction modification as discussed in the previous section. The abstract architecture can be mapped to a physical architecture to be used for the development of a DBMS using transaction modification or for the integration of this technique into an existing DBMS [Gref90c, Gref91b]. The latter case is shown in Chapter 6 for the PRISMA DBMS.

4.7.1 Architecture overview

The abstract DBMS architecture for constraint handling through transaction modification is shown in Figure 4.1. The transaction modification module, indicated by the $\oplus$ operator, forms the heart of the architecture. The following components can be identified in the architecture:

User Interface The User Interface forms the (interactive) interface to the system. The interface can both be used for data definition and data manipulation tasks.

DML Compiler The Data Manipulation Language compiler translates the language used for the specification of user transactions into extended relational algebra.

Query Optimizer The Query Optimizer transforms a user transaction into an equivalent optimized transaction.

Constraint Analyzer The Constraint Analyzer allows the user to define new or modify existing constraints, and performs the validation of these constraints.

Trigger Generator The Trigger Generator generates the trigger sets for new or modified constraints; this component implements the function *GentrigC*.

Data Dictionary The Data Dictionary is used for the storage of all meta-data of the system, like the database schema, defined integrity rules and derived integrity programs.

Constraint Optimizer The Constraint Optimizer performs the optimization of newly defined constraints; this component implements the function *OptR*.

Constraint Compiler The Constraint Compiler translates constraints from the integrity rule format to triggered programs format; this component implements the function *TransR*.

4.7 Abstract system architecture

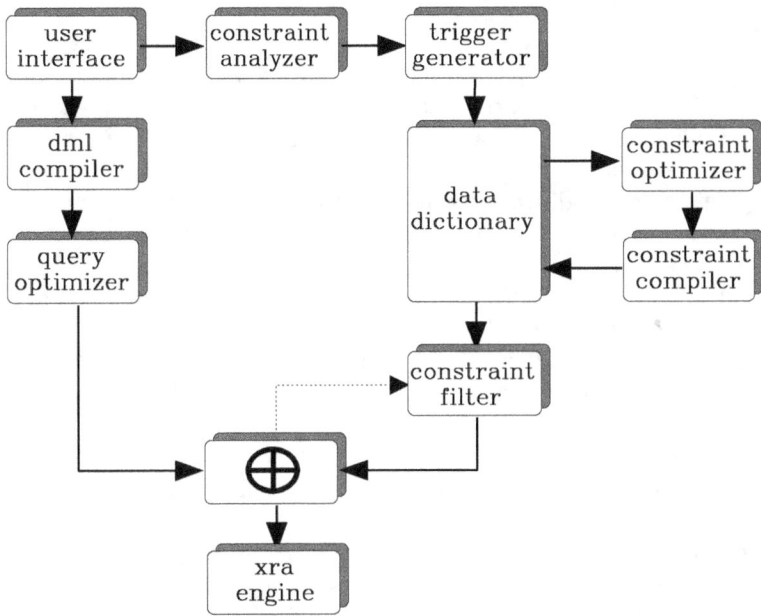

Figure 4.1: Abstract system architecure for transaction modification

Constraint Filter The Constraint Filter selects the constraints to be enforced at the end of a transaction; it implements the function *SelPS*.

Transaction Modification Module This module performs the actual transaction modification; this coincides with the top layer functionality of the *ModT* function.

Relational Engine The eXtended Relational Algebra Engine is responsible for the actual execution of a modified transaction including concurrency, atomicity and durability control.

One aspect of the abstract system architecture may need further explanation. As shown in Figure 4.1, the actual transaction modification takes place after the user transaction has been optimized by the Query Optimizer. Consequently, the transaction modifier is *not* processed by the Query Optimizer, and all constraint optimization is performed by the Constraint Optimizer. This design decision was taken for two main reasons. In the first place, the Query Optimizer should not be bothered with constraint-specific details, like alarm statements and differential sets. In the second place, in a situation with many constraints, the transaction modifier can be complex; consequently, static (definition time) optimization of constraints is preferable to keep transaction response times as low as possible.

4.7.2 Architecture behaviour

The User Interface component is used both for data definition and data manipulation purposes. In the context of this thesis, data definition is restricted to integrity constraint or rule definition. Data manipulation commands are submitted to the system in user transactions.

A newly defined or modified integrity constraint definition is first processed by the Constraint Analyzer module. If the constraint is found to be correct, the Trigger Generator generates the trigger set to obtain a complete integrity rule. This rule is stored in the Data Dictionary. Further, the rule is optimized and translated into an integrity program; this program is also stored in the Data Dictionary. The constraint in rule format is used for analysis, information and modification purposes; the constraint in integrity program format is used for constraint enforcement.

A user transaction is first translated into extended relational algebra by the DML Compiler, and optimized by the Query Optimizer component. Next, the transaction is modified by the Transaction Modification Module. This component makes use of the Constraint Filter to retrieve only the integrity programs associated with constraints that may be violated by the user transaction. The modified transaction is then executed in the normal way by the eXtended Relational Algebra engine.

Chapter 5

Fragmentation and parallelism

The previous chapter has discussed the transaction modification approach to integrity control in a centralized database system context. This chapter extends this technique to the context of distributed and parallel database systems.

The first section of this chapter discusses the general ideas behind distributed and parallel database systems by identifying the class of systems considered in this thesis, and presenting a short overview of the history and state of the art of these systems. The second section extends the database concepts of Chapter 2 to distributed databases; these concepts are used to describe databases that are managed by multiple processing units. Given the fact that users should not be bothered with the distribution of a database, integrity constraints are specified with respect to a centralized view of a distributed database. Techniques to transform these constraint definitions to distributed forms are discussed in Section 5.3. The last two sections of this chapter are devoted to optimization and translation of integrity rules, and enforcement of constraints; these sections present extensions of the ideas of the corresponding sections in the previous chapter.

The techniques described in this chapter are used in the PRISMA parallel database system, described in the next chapter.

5.1 Distribution and parallelism

This section presents the basic ideas and concepts of distribution and parallelism in database systems. First, the different types of distributed database systems are presented, and a choice is made for the type considered in this thesis. The history and state of the art of this type is presented next.

5.1.1 Distributed database systems

In general, distributed database systems are systems that use more than one processor for database management tasks. Two main types of distributed database systems can be distinguished, based on the reason for the use of multiple processors:

Geographically Distributed Systems Geographically distributed database systems are systems in which the processors are necessarily geographically distributed, because the use of the system requires database processing at multiple geographically distributed sites; a rationale for this type of system can be found in [Ceri84].

Parallel Systems Parallel database systems are systems in which multiple processors are used at the same site to obtain high system performance through the use of parallelism in data processing. The processors are either connected in a local area network, or combined into one integrated system, called a *parallel database machine*.

In geographically distributed systems, distribution is the starting point, and parallelism may be used if possible; in parallel systems, on the other hand, parallelism is the starting point, and distribution is required to obtain parallelism. This thesis is mainly concerned with parallel database systems; most of the techniques developed are applicable in geographically distributed systems as well, however.

Parallel database machines can be subdivided into two classes depending on the way database management functions are assigned to the processors in the system [Su88]. In systems with function replication, all processors can perform all or most of the database management functions. In systems with function distribution, each processor is assigned a specific part of the functions, either by programming general purpose processors differently, or by having different types of dedicated processors. Systems with replication of functions offer a number of advantages over systems with distribution of functions: extensibility, flexibility, availability, and reliability. Systems with replication of functions can be further subdivided into systems with static and systems with dynamic processor-memory interconnection [Su88]. Systems with dynamic processor-memory interconnection require special-purpose interconnection hardware, whereas systems with static interconnection can be based on standard hardware. Therefore, this type of system has received most attention in the recent past, and will be the hardware background for the work in this thesis; the PRISMA system discussed in the next chapter also belongs to this category. A short overview of history and state of the art of parallel database systems with functional replication and static processor-memory interconnection is presented below.

5.1.2 History and state of the art

The development of parallel database systems started in the late seventies. Early projects were the MICRONET project at the University of Florida, and the EDC project at the Electrotechnic Laboratory of Japan, both with a bus-structured interconnect [Su88]. The key feature of the MICRONET system is the use of custom-built bus and interfaces. The EDC multiprocessor makes use of magnetic bubble memories as secondary storage devices.

The DBC/1012 database computer built by Teradata Corporation in the early eighties is the only commercially available system in the category considered here [Su88]. The system is equipped with two intelligent interprocessor buses called Ynets, forming an essential feature in the design of the DBC/1012.

From the mid-eighties a number of research projects investigated the use of shared-nothing topologies built from standard hardware, i.e. networks consisting of standard processing nodes sharing nothing else but the network. The GAMMA system has been under development at the University of Wisconsin [DeWi90]. The earlier versions of the system were based on a number of mini-computers interconnected by a ring network. More recent versions are implemented on hardware with a hypercube architecture. In about the same period, the Bubba system was developed at MCC [Boral90]. This system aimed both at high performance through parallelism and complete functionality through the use of a sophisticated database programming language called FAD [Hart88]. A third project to be mentioned here is the HC16-186 project conducted at the Norwegian Institute of Technology [Brat89]; the HC16-186 machine and its successors have a hypercube topology in which dual-ported memories are used for data communication between pairs of processors. Finally, the PRISMA project discussed in the next chapter is a member of this generation.

Most parallel database system projects that have been completed by now focus heavily on performance-related issues; many even concentrate on parallel execution of single relational operations. The Bubba and PRISMA projects form exceptions here: Bubba also focusses on a complex data model and database programming language, and the PRISMA project is the only research effort in this category paying major attention to integrity control.

Projects that have started recently make use of more complex hardware architectures, e.g. networks in which each node is also a parallel system, and pay more attention to high-level functionality. An example of projects in this generation is the EDS project, an ESPRIT project by several European partners [Wats90, Skelt92] focussing both on high performance and complete functionality.

5.2 Distributed database concepts

This section discusses the basic concepts of distributed databases necessary for the treatment of integrity control in a parallel context. The discussion is partially based on the work in [Ceri84].

The hardware of a distributed database system consists of a number of nodes interconnected by means of a communication network. Each node is equipped with local data processing and data storage facilities. Nodes are identified with their node number $1, \ldots, n$, where n is the number of nodes in the network. Each node stores the software necessary to perform all or most of the database management functions. The data of a distributed database is spread over the nodes of a distributed database system. The way in which the data is split into parts is called the *data fragmentation*, the way the parts are assigned to the nodes in the system is called *data allocation*.

5.2.1 Data fragmentation and allocation

The fragmentation of a database determines how the global database is split up into multiple parts. Each global relation is fragmented into one or more parts, called *relation fragments* or simply *fragments*. In general, the fragmentation of a global relation should satisfy the following properties [Ceri84]:

- The fragmentation must be complete, meaning that all data of the relation must be included in the fragments of the relation.

- The fragmentation should be disjoint: there should be no duplication in the storage of the relation data.

- The global relation must me reconstructable from the fragments, i.e. the global relation can be derived from the fragments by means of algebraic operations.

Various forms of data fragmentation are described in the literature: range-based horizontal fragmentation, hash-based horizontal fragmentation, vertical fragmentation, and mixed fragmentation [Ceri84]. Here, we limit ourselves to hash-based horizontal fragmentation, since this form enables parallel query execution best. In the sequel, the term fragmentation refers to horizontal hash-based fragmentation.

Definition 5.1 A *data fragmentation schema* for relation schema $\mathcal{R}$ is a pair $FS_\mathcal{R} = \langle \alpha, d \rangle$, where α is a duplicate-free list of attributes of $\mathcal{R}$, and $d \in I\!\!N^+$; α is referred to as the *fragmentation attributes* and d as the *fragmentation degree* of $\mathcal{R}$. The schema $FS_\mathcal{R}$ defines a set of *relation fragments* $\{R_1, \ldots, R_d\}$ for an instance R of schema $\mathcal{R}$:

$$R_i = \sigma_{hash(\alpha) \bmod d = i-1} R \text{ for } 1 \leq i \leq d$$

Here, $hash$ is a system-defined function that maps an arbitary tuple to an integer, and mod denotes the integer modulo function [Gill76]. The set of fragment names of a relation R is denoted as $fragments(R)$, the fragmentation attributes of R as $\alpha_f(R)$, and the fragmentation degree of R as $\delta(R)$. □

5.2 Distributed database concepts

The fragmentation defined above assigns each tuple of a relation R to a relation fragment R_i, dependent on the value of the attributes in α. Clearly, this fragmentation strategy satisfies the fragmentation requirements, since each tuple of a relation R is assigned to a relation fragment, no tuple is assigned to more than one fragment, and relation R is obtained by taking the union of the fragments of R:

$$R = R_1 \cup \cdots \cup R_d$$

Each relation fragment R_i is mapped to a node of the distributed database system. This mapping is specified by a data allocation schema.

Definition 5.2 A *data allocation schema* for a relation schema $\mathcal{R}$ and its data fragmentation schema $\langle \alpha, d \rangle$ is a list $AS_\mathcal{R} = \langle a_1, \ldots, a_d \rangle$, where $a_i \in I\!N^+$. Schema $AS_\mathcal{R}$ defines a mapping of each fragment R_i of an instance R of schema $\mathcal{R}$ to a node a_i of a distributed database system. □

5.2.2 Distributed relations and databases

Given the fragmentation and allocation concepts introduced above, distributed relations and distributed databases can be defined as follows.

Definition 5.3 A *distributed relation schema* is a triple $\mathcal{R}^D = \langle \mathcal{R}, FS_\mathcal{R}, AS_\mathcal{R} \rangle$, where $\mathcal{R}$ is a relation schema as defined in Chapter 2, $FS_\mathcal{R}$ is a fragmentation schema, and $AS_\mathcal{R}$ an allocation schema. A *distributed relation instance* or *distributed relation* is a set of fragments defined by $FS_\mathcal{R}$ and allocated as specified by $AS_\mathcal{R}$. □

Definition 5.4 A *distributed database schema* $\mathcal{D}^D$ is a set of distributed relation schemas $\{\mathcal{R}_1^D, \ldots, \mathcal{R}_n^D\}$. A *distributed database instance* or simply *distributed database* is a set of distributed relation instances. □

A distributed database schema for the example beer database in Appendix B is shown in Table 5.1; this schema can be used for a distributed database system with at least eight nodes.

5.2.3 Fragmentation and allocation transparency

As described above, in distributed database systems relations are fragmented and allocated to various nodes in the system. For users and application programs, it is however much more convenient to work on a centralized version of the database. Therefore, a distributed database system can offer various levels of transparency [Ceri84].

$\mathcal{R}$	$attributes(\mathcal{R})$	$FS_\mathcal{R}$	$AS_\mathcal{R}$
beer	name, brewery, type, alcperc	$\langle\langle\text{name}\rangle, 4\rangle$	$\langle 5, 6, 7, 8\rangle$
brewery	name, city, country, part_of	$\langle\langle\text{name}\rangle, 1\rangle$	$\langle 1\rangle$
pub	name, city, country, brewery	$\langle\langle\text{city,country}\rangle, 3\rangle$	$\langle 2, 3, 4\rangle$
trade	pub, beer, qty_bought, qty_sold	$\langle\langle\text{beer}\rangle, 4\rangle$	$\langle 5, 6, 7, 8\rangle$

Table 5.1: Example distributed database schema

Definition 5.5 A database system offers *allocation transparency* if users and application programs can be fully unaware of the allocation of relation fragments. This means that users of the system 'see' a fragmented database as if it exists on a single node. □

Definition 5.6 A database system offers *fragmentation transparency* if users and application programs can be fully unaware of the fragmentation of relations. This means that users of the system 'see' a centralized database. □

Clearly, fragmentation transparency implies allocation transparency.

The fragmentation concept also applies to definitions of integrity constraints and rules: if a distributed database system offers fragmentation transparency with respect to integrity specification, integrity constraints and rules can be formulated in terms of global relations. Clearly, fragmentation transparency is a desirable concept in integrity specification for two main reasons:

- The database designer does not have to take fragmentation and allocation details into account; this simplifies integrity specification to a large extent.

- The specification of the integrity of the database does not require modifications by the database designer in case of a reorganization of fragmentation and/or allocation of the database.

To offer fragmentation transparency, a distributed database system must be equipped with mechanisms for this purpose; these mechanisms are described below.

5.3 Integrity rule fragmentation

As discussed above, a distributed database system should offer fragmentation transparency with respect to the definition of integrity rules. This requires techniques to translate rule definitions stated in terms of global relations into equivalent definitions stated in terms of relation fragments. This section describes this translation in two steps:

5.3 Integrity rule fragmentation

- The translation of an integrity rule stated in terms of global relations into a standard form stated in terms of relation fragments, called the *canonical fragment form*.

- The translation of an integrity rule in canonical fragment form into a set of integrity rules that take distributed constraint enforcement possibilities into account, called the *distributed fragment form* of the rule.

This translation is comparable with the translation of global queries into fragment queries as described in [Ceri84]. The techniques presented below can handle arbitrary fragmentation schemes and integrity rules. Other approaches may require restrictions on combinations of fragmentation schemes and rules [Ceri92].

The specification of constraints in terms of fragments requires an extension of the constraint specification language $\mathcal{CL}$ as introduced in Chapter 3. This extension is discussed below; next, the two translation steps of integrity rules are described.

5.3.1 Constraint language extension

According to the definition of the $\mathcal{CL}$ language given in Chapter 3, tuple variables in constraint definitions can vary only over relations or single relation fragments. To be able to easily specify constraints over reconstructions of global relations, we need to add a construct to the language that allows tuple variables to vary over the union of a set of fragments.

Definition 5.7 The constraint specification language $\mathcal{CL}$ is redefined as follows:

- The alphabet of the language is extended with the set of multi-set function symbols $FM = \{\cup\}$.

- The set of multi-set terms $\mathcal{T}_M$ is introduced; the following constructs are elements of this set:

    - A multi-set constant from the set M.
    - A multi-set function application $T_1 \cup T_2$, with $T_1, T_2 \in \mathcal{T}_M$.

- The set of terms $\mathcal{T}$ is renamed into the set of value terms $\mathcal{T}_V$; this set is redefined as follows:

    - A value constant from the set C.
    - An attribute selection $x.i$, where $x \in V$ and i an integer constant from C.
    - An arithmetic function application $t_1 \vartheta t_2$, where $\vartheta \in FV$, $t_1 \in \mathcal{T}_V$, and $t_2 \in \mathcal{T}_V$.
    - An aggregate function application $\Gamma(R, i)$, with $\Gamma \in FA$, $R \in \mathcal{T}_M$, and i an integer constant from C.

- A counting function application $\Gamma(R)$, with $\Gamma \in FC$ and $R \in T_M$.
- The set of atomic formulas $\mathcal{A}$ is redefined to consist of the following elements:
  - An arithmetic comparison $T_1 \vartheta T_2$, with $\vartheta \in PV$, and $T_1, T_2 \in T_V$.
  - A multi-set membership expression $x \in R$, where $x \in V$, and $R \in T_M$.
  - A tuple comparison $x = y$, where $x, y \in V$.

□

As may be clear, this extension of the $\mathcal{CL}$ language introduces two kinds of terms: value terms that have an atomic value from a basic domain, and multi-set terms that have a multi-set of tuples as value.

5.3.2 Canonical fragment form

The most straightforward fragmented form of an integrity rule is the canonical fragment form. This form is obtained by replacing the global relations in trigger set, condition, and action of the rule in the proper way by the fragments of these global relations. This procedure is defined formally by the algorithm below.

Algorithm 5.1 Let $J = \langle t, c, a \rangle$ denote an integrity rule stated in terms of global relations. Condition c is defined on relations $R1, R2, \ldots$, denoted as $c(R1, R2, \ldots)$. Now the canonical fragment form of J results from function $FragR$ shown below.

$FragR(J) = \langle FragT(t), FragC(c), FragA(a) \rangle$

$FragT(t) = \{\langle u, f \rangle \mid \langle u, r \rangle \in t \wedge f \in fragments(r)\}$

$FragC(c(R1, R2, \ldots)) = c\left(\bigcup_{i=1}^{\delta(R1)} R1_i, \bigcup_{i=1}^{\delta(R2)} R2_i, \ldots\right)$

$FragA(a) = \begin{cases} P_\varepsilon & \text{if } a = P_\varepsilon \\ FragS(head(a)) \oplus FragA(tail(a)) & \text{otherwise} \end{cases}$

▽

In this algorithm, function $FragT$ translates a trigger set defined on a set of relations into the equivalent trigger set defined on all fragments of the relations. Function $FragC$ translates a condition stated in terms of relations into an the equivalent condition stated in terms of the reconstructions of these relations. Function $FragA$ translates an extended relational algebra program stated in terms of global relations into an equivalent canonical program in terms of relation fragments by processing each of the statements in the program with function $FragS$, defined below.

5.3 Integrity rule fragmentation

Algorithm 5.2 Let s denote an extended relational algebra statement defined on relations $R1, R2, \ldots$, denoted as $s(R1, R2, \ldots)$. Now the canonical fragment form of s results from function *FragS* shown below.

$FragS(s(R1, R2, \ldots)) =$

$$\begin{cases} \begin{cases} V = e(\bigcup_{i=1}^{\delta(R1)} R1_i, \bigcup_{i=1}^{\delta(R2)} R2_i, \ldots); \\ delete(R1_1, V); \\ \vdots \\ delete(R1_{\delta(R1)}, V) \end{cases} & \text{if } s \text{ is} \\ & delete(R1, e(R1, R2, \ldots)) \\ \begin{cases} V = e(\bigcup_{i=1}^{\delta(R1)} R1_i, \bigcup_{i=1}^{\delta(R2)} R2_i, \ldots); \\ insert(R1_1, \sigma_{hfc(R1,1)} V); \\ \vdots \\ insert(R1_{\delta(R1)}, \sigma_{hfc(R1,\delta(R1))} V) \end{cases} & \text{if } s \text{ is} \\ & insert(R1, e(R1, R2, \ldots)) \\ \begin{cases} V = e(\bigcup_{i=1}^{\delta(R1)} R1_i, \bigcup_{i=1}^{\delta(R2)} R2_i, \ldots); \\ update(R1_1, V, \alpha); \\ \vdots \\ update(R1_{\delta(R1)}, V, \alpha) \end{cases} & \text{if } s \text{ is} \\ & update(R1, e(R1, R2, \ldots), \alpha) \\ V = e(\bigcup_{i=1}^{\delta(R1)} R1_i, \bigcup_{i=1}^{\delta(R2)} R2_i, \ldots) & \text{if } s \text{ is} \\ & V = e(R1, R2, \ldots) \\ ?e(\bigcup_{i=1}^{\delta(R1)} R1_i, \bigcup_{i=1}^{\delta(R2)} R2_i, \ldots) & \text{if } s \text{ is} \\ & ?e(R1, R2, \ldots) \end{cases}$$

where hfc is the hash fragmentation condition:

$hfc(R, i) = (hash(\alpha_f(R)) \bmod \delta(R) = i - 1)$

▽

In short, the functionality of function *FragS* is as follows. A modify statement on a relation is replaced by a list of equivalent modify statements on each of the fragments of the relation. In case of an insert statement, the multi-set to be inserted has to be divided among the fragments of the relation to be modified using the fragmentation of this relation. Further, relations in expression positions in all types of statements are replaced by the reconstructions of these relations from their fragments.

The above algorithms are illustrated with the two examples below. Note that rule optimizations are not used here for reasons of clarity.

Example 5.1 Take the following rule based on example constraint $I1$:

WHEN $INS(beer)$
IF NOT $(\forall x)(x \in beer \Rightarrow x.alcperc \geq 0)$
THEN $delete(beer, \sigma_{alcperc<0}(beer))$

Assuming that relation *beer* is fragmented into 4 fragments as shown in Table 5.1, the canonical fragment form of this rule is:

WHEN $INS(beer_1), INS(beer_2), INS(beer_3), INS(beer_4)$
IF NOT $(\forall x)(x \in (beer_1 \cup beer_2 \cup beer_3 \cup beer_4) \Rightarrow x.alcperc \geq 0)$
THEN $delete(beer_1, \sigma_{alcperc<0}(beer_1 \cup beer_2 \cup beer_3 \cup beer_4));$
$delete(beer_2, \sigma_{alcperc<0}(beer_1 \cup beer_2 \cup beer_3 \cup beer_4));$
$delete(beer_3, \sigma_{alcperc<0}(beer_1 \cup beer_2 \cup beer_3 \cup beer_4));$
$delete(beer_4, \sigma_{alcperc<0}(beer_1 \cup beer_2 \cup beer_3 \cup beer_4))$

□

Example 5.2 Take the following rule based on example constraint $I4$:

WHEN $INS(trade), DEL(beer)$
IF NOT $(\forall x)(x \in trade \Rightarrow (\exists y)(y \in beer \land x.beer = y.name))$
THEN *abort*

Assuming the fragmentation of the relations as shown in Table 5.1, the canonical fragment form of this rule is:

WHEN $INS(trade_1), INS(trade_2), INS(trade_3), INS(trade_4),$
$DEL(beer_1), DEL(beer_2), DEL(beer_3), DEL(beer_4)$
IF NOT $(\forall x)(x \in (trade_1 \cup trade_2 \cup trade_3 \cup trade_4) \Rightarrow$
$(\exists y)(y \in (beer_1 \cup beer_2 \cup beer_3 \cup beer_4) \land x.beer = y.name))$
THEN *abort*

□

5.3.3 Distributed fragment form

The canonical fragment form of an integrity rule as defined above is stated completely in terms of fragments and can thus be used for constraint enforcement on a distributed database. The semantics of the canonical rule form is still relation-oriented, however, since global relations are always reconstructed where data from the database is used, and updates to one involved relation fragment will always trigger constraint enforcement on all fragments of the relations. To avoid these inefeciencies, the distributed fragment form for integrity rules is introduced, which takes the separate fragments of the involved relations more into account.

5.3 Integrity rule fragmentation

Algorithm 5.3 Given an integrity rule $J = \langle t, c, a \rangle$ in canonical fragment form, the distributed fragment form of J is the set of rules resulting from function $DistR$:

$$DistR(J) = \{SimpR(J, f) \mid f \in TrigFrags(t)\}$$

$$TrigFrags(t) = \{f \mid \langle u, f \rangle \in t\}$$

$$SimpR(J, f) = \langle SimpT(t, f), SimpC(c, f), SimpA(a, f) \rangle$$

$$SimpT(t, f) = \{\langle u, f \rangle \mid \langle u, f \rangle \in t\}$$

$\triangledown$

In this definition, $SimpC$ and $SimpA$ are functions that simplify a condition respectively action with respect to a given fragment. These functions are in fact optimization functions: they affect the efficiency of the technique, not the functionality. Consequently, a base functionality can be chosen for these functions that can be extended as required. The simplest form for both functions is the identity function, simply returning its argument without modification. A simple though important form of the $SimpC$ function is given below.

Algorithm 5.4 Let $c(R_1 \cup \cdots \cup R_n)$ be a condition defined on the reconstruction $R_1 \cup \cdots \cup R_n$ of relation R. The function below simplifies the condition with respect to a fragment R_i of R:

$$SimpC(c(R_1 \cup \cdots \cup R_n), R_i) =$$

$$\begin{cases} c(R_i) & \text{if } c(R_1 \cup \cdots \cup R_n) \equiv (\forall x)(x \in (R_1 \cup \cdots \cup R_n) \Rightarrow W) \\ & \text{and } basesets(W) \cap \{R_1, \ldots, R_n\} = \emptyset \\ c(R_1 \cup \cdots \cup R_n) & \text{otherwise} \end{cases}$$

Here $basesets(W)$ denotes the set of the base fragments occurring in condition W.
$\triangledown$

The simplification of integrity rules is illustrated with the examples below.

Example 5.3 The distributed fragment form of the first example canonical rule shown above is a set of four similar rules $\mathcal{J} = \{J_1, J_2, J_3, J_4\}$. Each rule pertains to one fragment of relation *beer*, so for $1 \leq i \leq 4$:

$$\begin{aligned} J_i \; = \; & \text{WHEN } INS(beer_i) \\ & \text{IF NOT } (\forall x)(x \in beer_i \Rightarrow x.alcperc \geq 0) \\ & \text{THEN } delete(beer_i, \sigma_{alcperc<0} beer_i) \end{aligned}$$

$\square$

Example 5.4 The distributed fragment form of the second example canonical rule shown above is a set of eight rules $\mathcal{J} = \{J_1, \ldots, J_8\}$, since eight relation fragments are involved in the rule. Rules $J_1, \ldots, J_4$ are triggered by inserts to one of the fragments of relation *trade*. The conditions of these rules can be simplified using the algorithm presented above. So, for $1 \leq i \leq 4$ we have:

J_i = WHEN $INS(trade_i)$
IF NOT $(\forall x)(x \in trade_i \Rightarrow$
$(\exists y)(y \in (beer_1 \cup beer_2 \cup beer_3 \cup beer_4) \wedge x.beer = y.name))$
THEN *abort*

Rules $J_5, \ldots, J_8$ are triggered by delete operations to one of the fragments of relation *beer*. The conditions of these rules cannot be simplified, so for $5 \leq i \leq 8$ we have:

J_i = WHEN $DEL(beer_{i-4})$
IF NOT $(\forall x)(x \in (trade_1 \cup trade_2 \cup trade_3 \cup trade_4) \Rightarrow$
$(\exists y)(y \in (beer_1 \cup beer_2 \cup beer_3 \cup beer_4) \wedge x.beer = y.name))$
THEN *abort*

□

5.3.4 Transaction modification

Given the integrity rule fragmentation functions described above, we can construct a function that translates an entire set of integrity rules from the relation to the fragment form.

Algorithm 5.5 Given a set of integrity rules $\mathcal{J}$ defined in terms of global relations, the rule set fragmentation function *DisFrRS* computes the equivalent set of integrity rules defined in terms of relation fragments:

$$DisFrRS(\mathcal{J}) = \begin{cases} \emptyset & \text{if } \mathcal{J} = \emptyset \\ DistR(FragR(head(\mathcal{J}))) \cup DisFrRS(tail(\mathcal{J})) & \text{otherwise} \end{cases}$$

▽

Clearly, the *DisFrRS* function processes a set of rules by recursively traversing the set, and fragmenting and distributing each of the elements (rules). As described before, a set of integrity rules is interpreted as a list by imposing an arbitrary ordering on the elements of the set.

The *DisFrRS* function can easily be integrated into the transaction modification machinery described in Chapter 4. As shown below, the transaction modification algorithm is changed only by adding the rule set fragmentation function. In this algorithm, user transactions are assumed to be stated in terms of fragments, i.e. to have been processed already by a fragmentation algorithm.

Algorithm 5.6 The transaction modification function for a distributed database environment is as follows:

$ModT(T, \mathcal{J}) = ModP(T\downarrow, \mathcal{J})\uparrow$
where
$ModP(P, \mathcal{J}) = \begin{cases} P & \text{if } TrigP(P, \mathcal{J}) = P_\varepsilon \\ P \oplus ModP(TrigP(P, \mathcal{J}), \mathcal{J}) & \text{otherwise} \end{cases}$
and
$TrigP(P, \mathcal{J}) = TrOptRS(DisFrRS(SelRS(P, \mathcal{J})))$

▽

This transaction modification algorithm is identical to that developed in Chapter 4, except for the fact that the *DisFrRS* function has been added. Note that the *DisFrRS* function can be applied to a rule set when the rules are defined, thereby resulting the same algorithm optimization described for integrity rule optimization and translation in Chapter 4.

As will be clear from the above algorithm, the actual modification of a user transaction takes place after user transaction and transaction modifier have been converted from relation-based to fragment-based form. This design decision was taken for the following reasons. In the first place, the standard transaction execution mechanism should not be bothered with fragmentation algorithms for specific constraint-related constructs, like alarm statements. In the second place, static (definition time) fragmentation of constraints is preferred to obtain optimal system performance.

5.4 Integrity rule optimization and translation

In Chapter 4 the optimization and translation of integrity rules was discussed in a centralized database context. A distributed database context requires new extensions to the optimization and translation techniques, i.e. to the functions *TransR* and *OptR* introduced in Section 4.4. These extensions are discussed in this section.

5.4.1 Constraint optimization

As discussed in Chapter 4, in integrity rule optimization we focus on the optimization of the condition of integrity rules, i.e. on techniques to optimize $\mathcal{CL}$ constructs. In Section 4.4.1 a number of techniques have been discussed. Here, a few new techniques are added that are specific for constraints defined on the reconstruction of global relations. The presented techniques have strong similarities with those used in query optimization for distributed systems [Ceri84].

C	$OptC(C)$
$(\forall x \in \bigcup_{k=1}^{k=m} R_k)(c(x))$	$\bigwedge_{k=1}^{k=m}((\forall x \in R_k)(c(x)))$
$(\exists x \in \bigcup_{k=1}^{k=m} R_k)(c(x))$	$\bigvee_{k=1}^{k=m}((\exists x \in R_k)(c(x)))$

Table 5.2: Rewriting rules for distributed constraints

Syntactical manipulation

In query optimization it is very common to rewrite expressions to optimize the execution of the expression. For fragmented databases, one of the most frequently used techniques is pushing operations through unions to obtain computations that are local to the fragments [Ceri84].

A comparable approach can be applied to the conditions of integrity rules. Here we can push a quantifier through a union operator using the rewriting rules shown in Table 5.2. This optimization technique is illustrated by the example below.

Example 5.5 Take the following example rule used before:

WHEN $INS(trade_1)$
IF NOT $(\forall x)(x \in trade_1 \Rightarrow (\exists y)(y \in \bigcup_{i=1}^{i=4} beer_i \land x.beer = y.name))$
THEN $abort$

In the condition of this rule, variable y ranges over the reconstruction of global relation $beer$. The quantification can be made more local by pushing the $\exists$-quantifier through the union over the fragments of relation $beer$. As a result, the condition has the following form:

$(\forall x)(x \in trade_1 \Rightarrow \bigvee_{i=1}^{i=4}(\exists y)(y \in beer_i \land x.beer = y.name))$

□

Usage of fragmentation knowledge

Similar to the use of fragmentation knowledge in query optimization [Ceri84], fragmentation knowledge can be used for the optimization of constraints in systems with fragmented relations [Gref89a, Gref90a].

An important example is a referential integrity constraint between two horizontally fragmented relations. If the referring relation is fragmented on the referring attributes (foreign key) and the referred relation is fragmented on the referred attributes (key) using the same fragmentation algorithm, the enforcement of the

5.4 Integrity rule optimization and translation

referential integrity constraint can be reduced from a pairwise checking of all combinations of fragments of the two relations to a pairwise checking of the compatible fragments only.

Example 5.6 In the previous section the distributed fragment form of referential integrity constraint $I4$ of the example database has been derived. If the involved relations are fragmented on the key, respectively foreign key attributes, and have the same fragmentation degree (as shown in Table 5.1), the distributed fragment form can be optimized to the rule set $\mathcal{J} = \{J_1, \ldots, J_8\}$, where for $1 \leq i \leq 4$ and $5 \leq j \leq 8$:

$$J_i = \text{WHEN } INS(trade_i)$$
$$\text{IF NOT } (\forall x)(x \in trade_i \Rightarrow (\exists y)(y \in beer_i \land x.beer = y.name))$$
$$\text{THEN } abort$$

$$J_j = \text{WHEN } DEL(beer_{j-4})$$
$$\text{IF NOT } (\forall x)(x \in trade_{j-4} \Rightarrow (\exists y)(y \in beer_{j-4} \land x.beer = y.name))$$
$$\text{THEN } abort$$

□

A second important example for the usage of fragmentation knowledge is a tuple-oriented transition constraint, i.e. a constraint that specifies a condition on the old and new values of a tuple. In general, the old and new values of a tuple are not necessarily in the same relation fragment. Consequently, constraint enforcement over the global relation is necessary. If the relation is fragmented on the key attributes that identify a tuple, however, local enforcement per relation fragment is possible.

Example 5.7 We take transition constraint $I7$ of the example beer database as an example. The distributed fragment form of the integrity rule associated with this constraint is a set of rules, each of which has the following form:

WHEN $INS(beer_i)$
IF NOT $(\forall x, y)((x \in beer_i \land y \in \bigcup_{j=1}^{\delta(beer)} beer_{old,j} \land x.name = y.name) \Rightarrow$
$(x.alcperc \geq y.alcperc))$
THEN $abort$

The union in the condition of this rule is necessary, because it is unknown in which fragment the old value of a $beer$ tuple is located. If relation $beer$ is fragmented on the attribute $name$, identifying a $beer$ tuple, the above rule can be simplified to the following:

WHEN $INS(beer_i)$
IF NOT $(\forall x, y)((x \in beer_i \land y \in beer_{old,i} \land x.name = y.name) \Rightarrow$
$(x.alcperc \geq y.alcperc))$
THEN $abort$

□

5.4.2 Constraint translation

In Chapter 4 the translation of integrity constraints to extended relational algebra constructs has been discussed. Here we extend these ideas to constraints defined on fragmented relations, i.e. constraints containing the $\mathcal{CL}$ extensions discussed in Section 5.3.1. In the context of parallel database systems, fragmented relations are used to obtain possibilities for parallel data processing. Therefore, the translation of constraints to algebra should support parallel processing. For this reason, the extended relational algebra as defined in Chapter 2 and 4 is equipped with data distribution primitives that support parallel processing. These primitives are used next for the translation of distributed constraints to parallel extended relational algebra programs.

Data distribution primitives

To perform parallel data processing in a database system, data to be processed must be available at multiple processing nodes. Two situations can occur here: either the data is already spread over multiple nodes, or it is not and has to be distributed. For the last situation, two data distribution primitives are added to the extended relational algebra.

Definition 5.8 The *extended relational algebra statements* are redefined as follows. Let E denote an arbitrary extended relational algebra expression as defined in Definition 2.9. Then the following constructs are extended relational algebra statements:

- The extended relational algebra statements as defined in Definition 4.1.

- The *copy* statement $copy(E, T_1, \ldots, T_n)$ assigns the multi-set E to each of the new and implicitly defined relational variables T_i in a parallel fashion:

$$copy(E, T_1, \ldots, T_n) \equiv \begin{cases} T_1 = E, \\ T_2 = E, \\ \vdots \\ T_n = E \end{cases}$$

- The *split* statement $split(E, a, T_1, \ldots, T_n)$ distributes the elements of the multi-set E over the new and implicitly defined relational variables T_i in a parallel fashion, where the assignment of individual tuples t to one of the

5.4 Integrity rule optimization and translation

C	$TransC(C)$
$(\forall x \in X)\left(\bigvee_{i=1}^{n}(\exists y \in Y_i)(x.i = y.j)\right)$	$copy(unique(\pi_i X), T_1, \cdots, T_n);$ $alarm((T_1 - \pi_j Y_1) \cap \cdots \cap (T_n - \pi_j Y_n))$
$(\forall x \in X)\left(\bigwedge_{i=1}^{n}(\forall y \in Y_i)(x.i \neq y.j)\right)$	$copy(\pi_i X, T_1, \cdots, T_n);$ $alarm((T_1 \cap \pi_j Y_1) \cup \cdots \cup (T_n \cap \pi_j Y_n))$

Table 5.3: Translation of typical constraint constructs

variables is determined by the value $hash(t.a)$:

$$split(E, a, T_1, \ldots, T_n) \equiv \begin{cases} T_1 = \sigma_{hash(a) \bmod n = 0} E, \\ T_2 = \sigma_{hash(a) \bmod n = 1} E, \\ \vdots \\ T_n = \sigma_{hash(a) \bmod n = n-1} E \end{cases}$$

The punctuation by means of commas between statements in the extended relational algebra programs above is used to indicate simultaneous execution. □

Translating constraints to parallel algebra

Given the new extended relational algebra constructs introduced above, we can translate constraints over fragmented relations into parallel relational algebra programs. As discussed before, we concentrate on translating aborting integrity rules here.

Translation techniques for aborting integrity rules in a context without fragmented relations have been presented in Section 4.4.2. Translation techniques of typical constructs with fragmented relations are shown in Table 5.3.

Example 5.8 Take the following aborting integrity rule discussed before:

WHEN $INS(trade_1)$
IF NOT $(\forall x)(x \in trade_1 \Rightarrow \bigvee_{i=1}^{i=4}(\exists y)(y \in beer_i \land x.beer = y.name))$
THEN $abort$

This rule can be translated to the following extended relational algebra construct:

$copy(unique(\pi_{beer} trade_1), T_1, \ldots, T_4);$
$alarm((T_1 - \pi_{name} beer_1) \cap \ldots \cap (T_4 - \pi_{name} beer_4))$

If the *beer* relation is fragmented on the key attribute of the referential integrity constraint, the algebra construct above can be optimized. Using the *split* statement, the foreign key values can be 'directed' only to the corresponding fragment of relation *beer*, thereby reducing the left operand of the difference operations and completely avoiding the intersect operations:

$split(unique(\pi_{beer} trade_1), (\%1), T_1, \ldots, T_4);$
$alarm((T_1 - \pi_{name} beer_1) \cup \ldots \cup (T_4 - \pi_{name} beer_4))$

□

5.5 Constraint enforcement

As discussed in Chapter 4, in the transaction modification approach to integrity control constraint enforcement takes place through the execution of modified transactions. In a distributed database context, the same principle can be used, thus ensuring atomicity, serializability, and durability of transaction execution. The only difference here is the fact that modified transactions should be executed in a parallel fashion.

5.5.1 Parallelism in transaction execution

As integrity constraints are translated to extended relational algebra programs and are fully integrated into transactions, parallelism in constraint enforcement coincides with parallelism in regular transaction execution. Taking an integrity control point of view, we can make a classification of the possibilities for parallelism, however [Gref89a, Gref90a]:

Inter-Constraint Parallelism Inter-constraint parallelism is achieved when extended relational algebra programs associated with multiple integrity constraints are executed concurrently. We can distinguish between two forms here:

> **Inter-Transaction Parallelism** If the programs belong to different transactions, we use the term inter-transaction parallelism. This kind of parallelism exists if the system has parallel support for multi-user access to the database.
>
> **Intra-Transaction Parallelism** Inter-transaction parallelism means that multiple constraints are enforced for the same transaction. This kind of parallelism requires advanced scheduling techniques to be discussed in Chapter 7.

Intra-Constraint Parallelism Intra-constraint parallelism implies that the program associated with a single integrity constraint is executed in a parallel fashion. Here, parallelism exists in two forms:

5.5 Constraint enforcement

Horizontal Parallelism If one relational operation on a global relation is distributed over the fragments of this relation, the operation can be executed on the fragments in parallel. This is called horizontal parallelism or *task spreading*.

Vertical Parallelism If the enforcement of a constraint consists of multiple relational operations that have an input-output relationship, these operations can be executed in parallel. This kind of parallism is called vertical parallelism or *pipelining*.

Inter-constraint parallelism is easy to understand, since the constraints to be enforced are fully independent, so the associated relational algebra constructs have no relation either. Intra-constraint parallelism is illustrated by the example below.

Example 5.9 Take the following extended relational algebra program associated with a single integrity constraint at the fragment level of a database:

$copy(unique(\pi_{beer} trade_1), T_1, \ldots, T_4);$
$alarm((T_1 - \pi_{name} beer_1) \cap \ldots \cap (T_4 - \pi_{name} beer_4))$

Horizontal parallelism can be obtained here if the difference operators on the various fragments of relation *beer* are executed in parallel. Vertical parallelism exists if the copy, difference, and intersect operators are executed in parallel. □

The subject of parallelism is discussed in more detail in the next chapter, where it is placed in the context of a real-world parallel database system.

Chapter 6

Integrity control in PRISMA/DB

The previous chapters of this thesis have discussed concepts and algorithms for integrity control through transaction modification in a parallel database system. This chapter is devoted to the application of these techniques in a real-world parallel database system, called PRISMA/DB. Two major prototype versions of this system have actually been built and tested on parallel hardware. The second version of the system has been equipped with an integrity control subsystem using the transaction modification technique. This system has been used as a testbed for performance evaluation of integrity control.

This chapter starts with an introduction to the PRISMA project to give the reader an impression of the project context in which the practical part of the work presented in this thesis has been performed. The second section describes the PRISMA/DB database management system; both the design principles and the base architecture of the system are discussed. The chapter continues describing how the transaction modification technique has been integrated into the base architecture of PRISMA/DB. Next, a performance evaluation of parallel integrity control is presented, conducted on multi-processor hardware. The chapter ends with a few conclusions regarding implementation and tests.

6.1 Introduction to the PRISMA project

The PRISMA project was initiated in 1986 as a joint research effort between Philips Research Laboratories at Eindhoven, the Netherlands, and several Dutch academic institutions, among which are the University of Twente, the Center for Mathematics and Computer Science at Amsterdam, and the University of Amsterdam [Kers87, Apers88]. The project consisted of three main sub-projects:

Parallel Machine The machine subproject had two main goals. The first goal was the design and construction of a shared-nothing multi-processor computer called POOMA, consisting of a large number of processor nodes interconnected by means of a message-passing point-to-point network. The second goal was the design and efficient implementation of a parallel object-oriented programming language called POOL-X, that was to be used as the implementation language by the other two subprojects.

Database System The database system subproject aimed at the development of a parallel main-memory relational database system, called PRISMA/DB. PRISMA/DB was to have a flexible architecture to enable experimenting with functionality and performance in a parallel context. The experiments should result in a prototype database system with both high performance and complete functionality.

Expert System The goal of the expert system subproject was the design and implementation of an expert system shell that uses parallelism for high performance inferencing.

These three subprojects account for the name PRISMA, which is an acronym for **Pa**Rallel **I**nference and **S**torage **MA**chine. This chapter is mainly concerned with the database system subproject; results of the machine subproject are discussed where necessary, since the database system is implemented in POOL-X to run on the POOMA system. The expert system subproject is not relevant here.

Design and implementation of PRISMA/DB are described in detail in Section 6.2. To give the reader a background on the implementation platform of PRISMA/DB, a short introduction to the POOMA hardware architecture and the POOL-X programming language is given below. Further details on the PRISMA project can be found in [Amer90].

6.1.1 POOMA machine

The **P**arallel **O**bject-**O**riented **MA**chine is a multi-processor system consisting of an arbitrary number of nodes interconnected by means of a message-passing network. The network topology can be an arbitrary graph of degree 4, i.e. a graph in which each vertex has four outgoing edges. The system has a shared-nothing architecture, meaning that the nodes share nothing but the network. An example configuration is shown in Figure 6.1; this system consists of 16 processing nodes connected together in a 4 × 4 torus network topology [Vlot90]. Other possible network topologies are chordal rings, mesh topologies, and hypercubes.

The configuration of one POOMA node is shown in the right hand part of Figure 6.1. A node consists of a standard data processor (68020 CPU), a dedicated communication processor, 16 Mbyte of standard main memory (RAM), and a stable storage device (a magnetic disk). The communication processor has four

6.1 Introduction to the PRISMA project

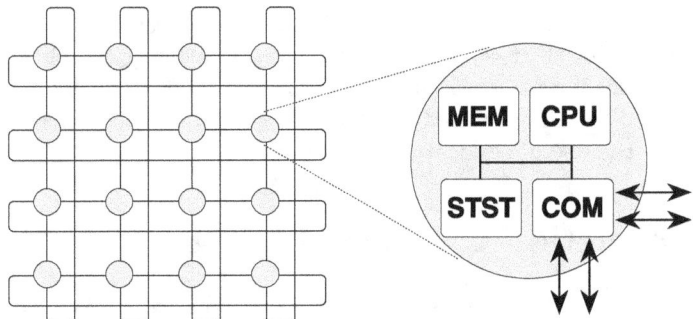

Figure 6.1: POOMA system and node architecture

bidirectional links for connections to other nodes. A node may be equipped with an ethernet interface to allow communication to a host machine.

Several POOMA configurations have been realized. The largest configuration consists of 100 nodes, half of which are equipped with a disk. This gives a total of 100 processors with a total integer performance of 200 MIPS, 1.6 Gbyte of main memory, and 15 Gbyte of disk storage [Vlot90]. Smaller configurations of 4 and 8 nodes are available.

6.1.2 POOL-X language

POOL-X is a **P**arallel **O**bject-**O**riented programming **L**anguage with e**X**tensions designed for the implementation of a database management system [Amer89, Spek90].

A POOL-X program consists of a number of *classes*, each of which contains definitions of data and operations on these data, called *methods*. A distinction is made between active and passive classes: active classes have a *body* that describes a process, passive classes do not have a process of their own. Classes are instantiated into *objects*. Objects of active classes form the processes of a parallel system; objects of passive classes are mainly used for data storage. Objects are created dynamically. Objects are also allocated dynamically to nodes of a multi-processor, either implicitly by the POOL-X run time system or explicitly by the application. The dynamic creation and allocation of active objects is a simple though powerful way to control parallelism in an application.

POOL-X objects communicate by means of message passing using their methods. Both synchronous and asynchronous methods are available: with synchronous message passing the sender has to wait for the receiver to answer a message, with asynchronous message passing, there is no answer and the sender does not have to wait. The allocation of objects on the various nodes of a system is transparent to the POOL-X communication.

To accomodate easy implementation of a database management system, POOL-X features tuple types with dynamic typing, i.e. tuple types can be created and manipulated run-time. These tuple types can be used directly for the representation of database tuples. Run-time compilation of operations on tuples is provided to obtain high efficiency in relational operations. Further, a stable storage I/O-system is provided at the POOL-X language level; this system provides storage stability through the use of replicated disk storage.

6.2 Introduction to PRISMA/DB

As mentioned above, PRISMA/DB is the database management system developed in the PRISMA project. The system was designed and implemented at the University of Twente and the Center for Mathematics and Computer Science from 1986 until 1991. The first prototype system, called PRISMA/DB0, was completed in 1989 [Gref89b]. The second prototype, PRISMA/DB1, was released in early 1991 [Gref91a].

This section discusses the design objectives, overall base architecture, and possibilities for parallelism in PRISMA/DB. Note that the base architecture design for PRISMA/DB does not include any facilities for integrity control. The next section discusses the integrity control subsystem added to the base architecture.

6.2.1 PRISMA/DB design

PRISMA/DB was designed to be a complete parallel relational database management system. The main design objectives were the following:

Performance The system should offer high performance data processing, especially in a context with large databases and complex operations.

Functionality The system should offer complete DBMS functionality, including several user interfaces, concurrency control, and crash recovery.

Flexibility The system should offer flexibility both in static architecture and in dynamic behaviour to enable experimenting with performance and functionality.

High performance data processing in PRISMA/DB is achieved through the use of parallelism in data processing and main memory storage of the entire database. To be able to easily use parallelism in data processing, the system was designed to work with a horizontally fragmented database. To allow maximum freedom in data allocation, fragmentation degree and fragment allocation can be chosen freely per relation by the database designer. The main-memory storage of the database avoids access to slow secondary storage for data retrieval. Note, however, that the POOMA hardware is not equipped with stable main-memory. Consequently, updates to the database have to be written to stable secondary storage to enable recovery from system crashes.

6.2 Introduction to PRISMA/DB

PRISMA/DB includes interfaces for four user data manipulation and data definition languages: standard SQL, a logic query language called PRISMAlog, an extended relational algebra language called XRA, and a dedicated data definition language called DDL [Gref91a]. The SQL interface is meant for standard applications, the PRISMAlog interface for deductive database applications including recursive queries, and the DDL interface for specification of data fragmentation and allocation. The data manipulation interfaces all offer complete fragmentation and allocation transparency. The system includes full support for concurrent access to the database by an arbitrary number of users. Crash recovery is included to be able to survive power and storage media failures.

Flexibility in PRISMA/DB was achieved through a strictly modular design of the system, with high level interfaces between the modules. Function separation is used as the basis for modularization. The modules are assigned specific database management tasks, like query optimization, transaction management, or concurrency control. Modules can be instantiated dynamically whenever needed. Where possible, the extended relational algebra language XRA is used as a high-level interface language between system modules.

These design decisions have all heavily influenced the global architecture and functionality of the architecture components of PRISMA/DB; this is discussed below.

6.2.2 PRISMA/DB base architecture

Figure 6.2 presents an overview of the global architecture of PRISMA/DB. This architecture consists of a highly dynamic process structure designed to manage a fragmented main-memory database in a parallel fashion.

Two central components of the system are the *data dictionary* (DD) and the *concurrency control unit* (CC). The data dictionary is the central storage of all meta-data of the system, like relation and fragment definitions. The component is used by most other components; therefore, these interfaces are not shown in the figure. The DD includes an interface for the XDD data definition language, used by the user interface component. The concurrency control unit controls concurrent access to the database. It uses a standard two-phase locking protocol with shared and exclusive locks, and relation fragments as the locking granularity. Further, it is equipped with a deadlock prevention algorithm [Elmas89].

The *user interface* (UI) forms the interactive interface to the system, used for data definition and data manipulation commands. The query preprocessing layer of the system is formed by the *query compiler* (QC) and *query optimizer* (QO) components. These components are instantiated once for each user session. The query compiler translates queries from the user data manipulation language into the internal relational language of the system (XRA); the type of query compiler is determined by the user language (SQL, PRISMAlog, or XRA). The QC component offers full fragmentation and allocation transparency to the user [Ceri84]. The query optimizer is responsible for the optimization of queries and the removal of

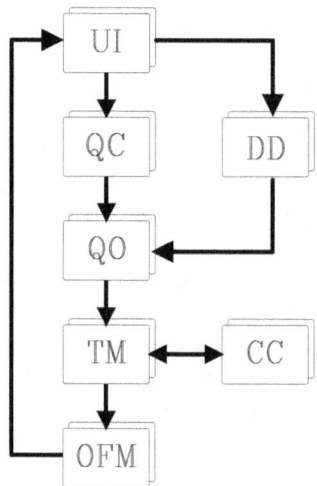

Figure 6.2: PRISMA/DB base architecture

fragmentation transparency. Further, this component transforms recursive queries into forms that can be handled by the lower layers of the system. The query optimizer uses a rewriting technique to perform its tasks.

The *transaction manager* (TM) forms the execution control layer of the system. This component is instantiated once for each transaction to be executed by the system. The TM is responsible for the creation of the necessary transient execution infrastructure for queries and the passing of commands to the execution layer of the system. Further, the TM controls the serializability of the transaction through a locking protocol in cooperation with the CC, and transaction atomicity through a two-phase commit protocol with the OFM layer of the system.

The *one-fragment manager* (OFM) component forms the data storage and query execution layer of the system. One-fragment managers come in two kinds: permanent and transient ones. Permanent OFM's are instantiated once for each base fragment of the database, and live as long as the relation exists to which the fragment belongs. All data of the database is kept in main memory to ensure fast retrieval. To ensure stability of the database over system crashes, updates are written to secondary storage using a logging and checkpointing technique. Permanent OFM's can also perform operations on the base data they manage. A permanent OFM can have several *local transaction manager* modules (LTM) that perform interface tasks to multiple concurrent transactions accessing the OFM. Transient OFM's are created by a transaction manager for the execution of relational operations and the storage of intermediate results of queries. The lifecycle of a transient OFM is a transaction. A transient OFM always has one LTM module, since it is used exclusively by the transaction it was created for. OFM's can exchange data through *channels*, which are abstract tuple transport links between

6.2 Introduction to PRISMA/DB

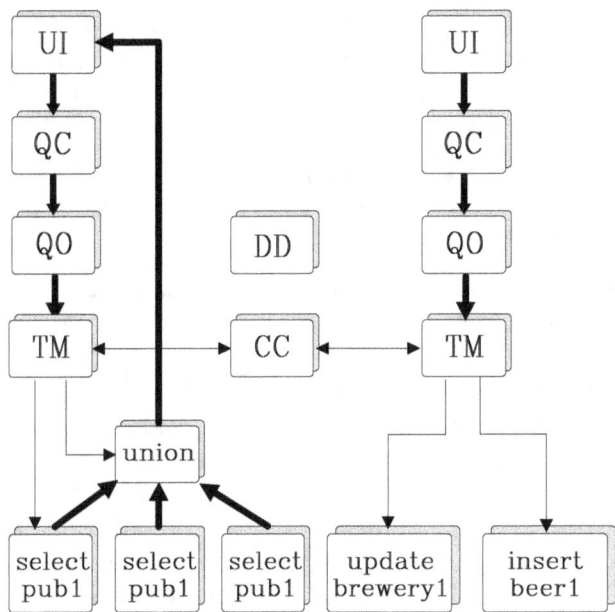

Figure 6.3: PRISMA/DB in execution

pairs of OFM's.

Figure 6.3 shows PRISMA/DB in action on the example beer database. Two concurrent user sessions are active. The left hand session is executing a transaction that performs a selection on the *pub* relation. A transient OFM has been created to perform the union on the results of the distributed selection. The right hand session is executing a transaction that performs an update on relation *brewery* and an insert on a fragment of relation *beer*.

6.2.3 Parallelism in PRISMA/DB

The architecture of PRISMA/DB is designed to have a broad range of possibilities for parallelism in data processing. The following forms of parallelism can be distinguished, all of which are illustrated in Figure 6.3:

Inter-transaction Parallelism The coarsest grainsize of parallelism applied in PRISMA/DB is parallelism between concurrently executing transactions. In Figure 6.3, two Transaction Managers are controlling two transactions in parallel.

Intra-transaction Parallelism Intra-transaction parallelism is paralellism within the execution of a transaction. Several forms can be distinguished here:

Inter-layer parallelism Several layers (components) of PRISMA/DB can work on one transaction in parallel: translation, optimization, control, and execution can be performed in parallel on multiple successive statements in one transaction.

Data layer parallelism Data-layer parallelism is parallelism within the data management layer (OFM layer) of PRISMA/DB. This form can again be subdivided as follows:

Inter-command parallelism Inter-command parallelism exists when multiple commands belonging to the same transaction are executed in parallel. In Figure 6.3, the right hand side transaction is performing an update and an insert statement in parallel.

Intra-command parallelism Intra-command parallelism exists when multiple operators within one command are executed in parallel. This type of parallelism can exist in two forms:

Horizontal parallelism Horizontal parallelism exists when one global relational operator is executed in parallel on multiple fragments of a relation or intermediate result. In Figure 6.3, horizontal parallelism is used in the execution of the global select operation on relation *pub*. This type of parallelism is also referred to as *task spreading*.

Vertical parallelism Vertical parallelism exists if two relational operators are executed in parallel that have a producer-consumer (operand-operator) relationship with respect to the data to be processed. In Figure 6.3, vertical parallelism exists between the execution of the select operations on the fragments of relation *pub* and the union operator that combines the results of the select operators. Vertical parallelism is also called *pipelining*.

The intra-command parallelism in PRISMA/DB is based on a data flow query execution technique [Wils90, Wils91].

6.3 Integrity control in PRISMA/DB

The base architecture of PRISMA/DB as discussed above is not equipped with any facilities for integrity control. This section discusses how an integrity control susbsystem is integrated into PRISMA/DB. This subsystem is a prototype implementation of the transaction modification technique as discussed in Chapters 4 and 5. The functionality of the prototype is restricted to aborting integrity rules based on a limited set of constraint types [Gref90e]. The restrictions were mainly made to avoid an excessive implementation effort; extensions to a subsystem with more complete functionality are possible.

6.3 Integrity control in PRISMA/DB

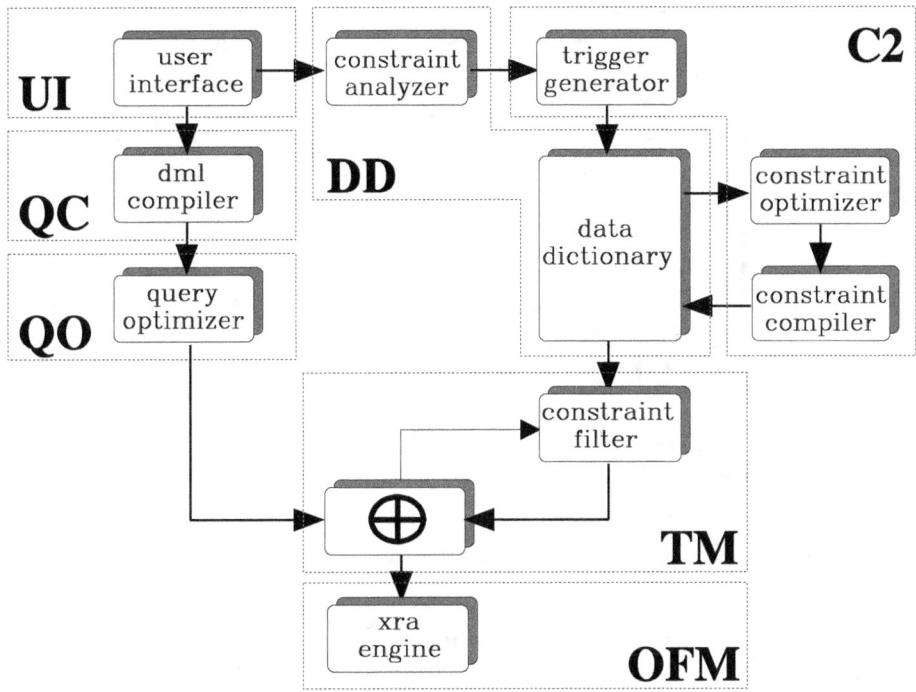

Figure 6.4: Abstract architecture in PRISMA/DB

6.3.1 Global architecture

To extend PRISMA/DB with constraint handling facilities, the abstract system architecture for transaction modification as depicted in Figure 4.1 is integrated into the PRISMA/DB base architecture. The modular structure of the abstract architecture enables an easy integration into the existing DBMS architecture. Figure 6.4 shows how the various tasks of the abstract architecture are allocated to PRISMA/DB components. Most tasks are integrated into existing components; one new component is introduced: the *Constraint Compiler* (C2). The extended architecture of PRSMA/DB is shown in Figure 6.5. The changes with respect to the base architecure are discussed below.

6.3.2 Constraint compiler

The Constraint Compiler (C2) is a new DBMS component. The external interface of the component is very simple, as one might expect from a compiler. The Constraint Compiler acts as a slave component for the Data Dictionary; the DD sends constraints to be compiled and receives the compiled constraints as result. The C2 component has the following functionality:

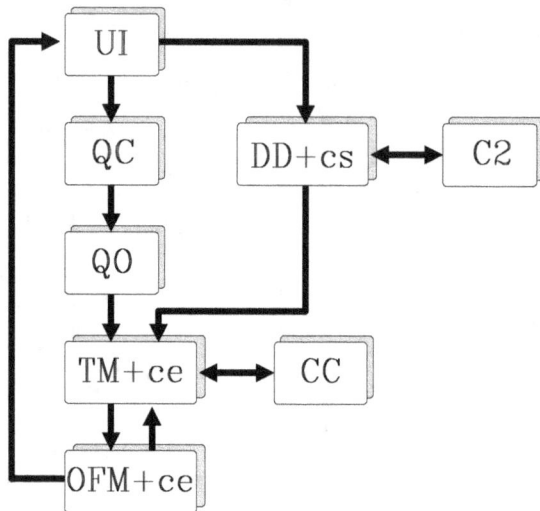

Figure 6.5: PRISMA/DB architecture with integrity control

- translation of constraints from relation to fragment level; as such, the C2 resolves fragmentation transparancy for constraint definitions;
- optimization of constraints;
- translation of constraints to XRA constructs; these constructs are directly executable by the Transaction Manager.
- generation of trigger sets for that are added to the translated constraints.

In the current prototype implementation of the PRISMA/DB Constraint Compiler, only a limited set of aborting integrity rules is accepted: domain and nonnull constraints, uniqueness (key) constraints, and referential integrity constraints (the structural constraint types of the relational data model [Gard89]). For reasons of easy implementation, the constraints are specified in the XDD language [Gref91a].

6.3.3 Data dictionary

The Data Dictionary is extended to be able to handle constraints. It stores constraints both in source format as specified by the user and in integrity program format (in XRA) as discussed in Chapter 4. The source format is necessary for automatic recompilation; the integrity program format is the form actually needed for constraint enforcement. Integrity programs are retrievable from the DD based on the relation fragments in their trigger sets.

The Data Dictionary automatically activates the Constraint Compiler when necessary; this occurs in the following cases:

6.3 Integrity control in PRISMA/DB

- definition of new constraints by the user;
- change in fragmentation of relations used in constraint definitions.

6.3.4 Transaction Manager

The Transaction Manager (TM) is extended to handle constraint filtering and transaction modification in the constraint enforcement process. The following tasks are performed for these purposes:

- During transaction execution, the TM analyzes the incoming XRA statements to record possible constraint violations per fragment; this is used to perform a first phase of constraint filtering based on fragment names.

- When the TM reaches end of transaction, it retrieves the constraint definitions for the possibly inconsistent fragments from the Data Dictionary.

- Next, the TM performs a second phase of filtering based on update types. The filtering algorithm uses the trigger sets generated by the Constraint Compiler.

- Finally, the TM performs the actual transaction modification. The constraint definitions resulting from the filtering phase are concatenated to the transaction.

The modified transaction is executed in exactly the same way as an unmodified transaction. This ensures that the execution of the transaction modifier conforms to serializability and atomicity of transactions. Further, all parallelizing algorithms as used for normal query processing are also available for constraint enforcement.

The TM tasks in constraint enforcement start when the TM reaches the end of transaction in the transaction specification, not in the transaction execution. This allows the TM to do its job during actual transaction execution, thereby reducing response times in parallel systems [Gref89a, Gref90b]. A transaction abort (either system or user generated) will cause the TM to suppress all further work on both normal transaction execution and constraint enforcement; this mechanism avoids unnecessary work as much as possible.

Constraint violations are handled using the same abort protocols that are also used to deal with other transaction failures (like media failure or division by zero).

6.3.5 One-Fragment Manager

Due to the fact that the transaction modification technique heavily relies on standard query operators for constraint enforcement, the modifications to the data management layer of PRISMA/DB are limited. Two extensions to the OFM component are necessary:

- In the first place, the OFM must support the alarm operator as discussed in Section 4.4.2. Due to the simple functionality of this operator, the implementation can be rather trivial. An abort situation can be handled by the existing abort mechanism.

- In the second place, the OFM must maintain the differential sets described in Section 4.4.1. The following differential sets are maintained by the OFM:

  insert set This tuple set contains all new tuples created by the current transaction, i.e. inserted tuples and new values of updated tuples.

  delete set The *delete* set contains all old tuples created by the current transaction, i.e. deleted tuples and old values of updated tuples.

  unchanged set This tuple set contains all tuples that have not been changed by the current transaction; in fact, this is not truly a differential set, but its existence allows further optimization of certain constraint types.

Further extensions to the OFM can be used to improve the performance of constraint enforcement; these extensions will be discussed in Chapter 8.

6.4 Performance evaluation

One of the key problems impeding the general use of integrity control in practice is formed by the high processing costs associated with integrity constraint enforcement on large databases. Notwithstanding this observation, little attention has been paid in literature to the performance analysis and evaluation of integrity control mechanisms in database systems. A short overview of the research in this field is presented below.

In [Badal79] a quantitative analysis of constraint enforcement is made, in which transaction compile time enforcement, transaction run time enforcement, and post transaction execution enforcement are compared with respect to the number of database accesses only. This and other simplifications limit the practical usability of the proposed model. In the context of the SABRE project some attention has been devoted to the performance modeling of constraint enforcement [Simon84, Simon85]. The cost of constraint enforcement is modeled as the number of I/O operations necessary to read and write data from and to disk. The only known work in the field of actual performance evaluation of existing constraint enforcement systems is also performed in the context of the SABRE project [Simon84, Simon85]. In these evaluations, however, database sizes are fairly small and response times are far from those of modern high-performance database systems. Recent research projects investigating the implementation of integrity control subsystems, like POSTGRES [Stone88, Stone90a] and Starburst [Widom91, Lohm91], do pay attention to efficiency aspects but have not yet published any work on the actual evaluation of the performance of integrity control.

6.4 Performance evaluation

This section presents a concise performance evaluation of the transaction modification technique as implemented in the PRISMA/DB parallel database system. First, the design of the benchmark and the setup of the experimental situation are discussed. Next, the results of the measurements on PRISMA/DB are presented. This section ends with a discussion of these results. Further details on the work in this section can be found in [Gref92b].

6.4.1 Benchmark design

When designing a performance evaluation for constraint enforcement in a parallel database system, one is confronted with a large number of parameters that have to be given a value. Important parameters are the types of the involved relations, sizes of the relations, fragmentation and allocation of the relations, types and definitions of constraints, number of operations in update transactions, type of update operations, selectivity of update operations, success rate of transactions, and level of concurrency. Further, there exist no commonly accepted benchmarks for this type of evaluation, like the Wisconsin benchmark for simple query execution [Bitt83]. Therefore, some choices have to be made, which may appear rather arbitrary. One of the main objectives in making choices has to be simplicity, to keep the experiments manageable and the results interpretable.

The following general choices were made for the measurements presented in this paper. The relevant quantity to be measured is *transaction response time*, i.e. the complete execution time of a transaction, including lock requesting, logging and checkpointing of updates to disk, and commit protocol. For reasons of simplicity, only single user situations are analyzed.

With respect to constraint types, three situations are analyzed: no constraint, domain constraint, and referential integrity constraint. For the referential integrity constraint, the extended form suggested in [Date81, Date90a] is used[1]. Note that the 'no constraint' situation means that constraint enforcement is enabled, but no constraints are defined. The main relation type is taken from the Wisconsin benchmark [Bitt83] and consists of 16 attributes (13 integer and 3 string attributes). To be able to experiment with relations with a larger cardinality, a simpler relation type of two integer attributes is used as well; this relation type is referred to as *Numbers*. Also for reasons of simplicity, transactions are taken to consist of one global insert operation only, where the tuples to insert are stored in a separate relation. The size of the relation to insert is 10% of the relation in which the insert takes place to model an update selectivity of 10%. Further, all transactions are designed not to violate any constraints. This has no consequence for the cost of the constraint enforcement processes. Transactions that are aborted due to constraint violation, however, can suppress their logging or checkpointing activity, since no changes are made to the database.

---

[1] As shown in the constraint definition in Table 6.1, the constraint includes null value semantics for the foreign key attribute.

rel.	type	card.	frag.	alloc.
W1	Wisconsin	10000	1..6	2..7
W2	Wisconsin	1000	1	8
W3	Wisconsin	1000	1..6	2..7
N1	Numbers	50000	1..6	2..7
N2	Numbers	5000	1	8
N3	Numbers	5000	1..6	2..7

tr.	definition
T1	begin
	insert(W1,W2)
	commit
T2	begin
	insert(N1,N2)
	commit

con.	type	definition
I1	domain	$(\forall x)(x \in R1 \Rightarrow x.a \geq 0)$
I2	ref. int.	$(\forall x)((x \in R1 \wedge x.a \neq null) \Rightarrow (\exists y)(y \in R3 \wedge x.a = y.b))$

Table 6.1: Benchmark relations, constraints, and transactions

An overview of the benchmark characteristics is given in Table 6.1. For each relation the type, cardinality, number of fragments, and allocation of the fragments on the processor nodes is shown.

6.4.2 Experimental situation

Above, the benchmark characteristics for the performance evaluation of constraint enforcement have been presented. Apart from these, the exact experimental situation has to be defined. The following choices are made:

- The measurements investigate the relationship between transaction response time and the degree of parallelism in the data management layer of PRISMA. Since PRISMA is a main-memory DBMS, attention will also be paid to the effect of main-memory data storage on constraint enforcement.

- The response times to be measured are complete transaction execution times measured at the transaction manager level of the system, i.e. the real time between transaction manager startup and transaction end. All transaction overhead, like concurrency control and logging, is taken into account to measure a realistic situation. Query translation and optimization are not taken into account, because the PRISMA/DB query compiler and optimizer components have not yet been properly optimized, and would thus obscure the measurement results.

- The measurements are performed on an 8-node POOMA machine. Although a larger configuration exists, the 8-node machine will be large enough to get a good impression of parallelism in constraint enforcement, without getting a situation too complex to analyze.

6.4 Performance evaluation

- The transaction manager process runs on a private node. Six nodes are used for the relation to be modified (also the relation on which the constraints are defined), and an auxiliary relation for the referential integrity constraint. The last node of 8 is used for the data to be inserted into the relation to be modified.

6.4.3 Measurements results

This subsection presents the measurement results of the performance evaluation described above. First, the measurements to establish the effect of parallelism on constraint enforcement are presented. Next, attention is paid to the measurements related to the effect of main-memory data storage. A discussion of the results is presented in the next subsection.

The effect of parallelism

Figure 6.6 shows the measurement results of the execution of transaction $T1$ on the *Wisconsin* relations (see Table 6.1). Four situations are shown:

1. no constraint.

2. domain constraint $I1$.

3. referential integrity constraint $I2$ with relations $W1$ and $W3$ fragmented and allocated compatibly with respect to the constraint.

4. referential integrity constraint $I2$ with relations $W1$ and $W2$ fragmented incompatibly with respect to the constraint.

In situation 3, matching tuples of the key and foreign key relations are always allocated on the same node; consequently, no inter-node data transport is necessary for constraint enforcement. In situation 4, foreign key values have to be transported to other nodes.

Figure 6.6 clearly shows that all four situations benefit from the use of parallelism. A number of remarks can be made, though. In the case without constraints, parallelism is effective up to 4 processors. This is caused by relatively high protocol overhead between the transaction manager and the data manager. The enforcement of domain constraint $I1$ is cheap compared to the execution of the transaction without constraint enforcement; here, not much can be gained from parallelism in constraint enforcement. The case of referential integrity constraint $I2$ with compatible fragmentation shows good use of parallelism: both the complete transaction execution time and the constraint enforcement overhead are reduced by parallelism. The optimal number of processors is greater than 6; further experiments have to be conducted here to establish this number. The case with non-matching fragmentation shows the same behaviour up to 3 processors;

6. Integrity control in PRISMA/DB

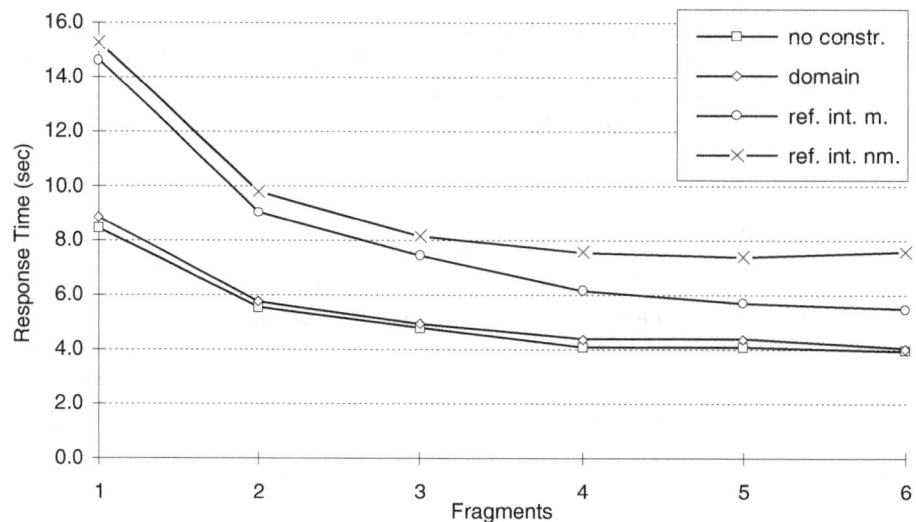

Figure 6.6: Speedup Wisconsin benchmark

with a larger number of processors, the transaction manager overhead dominates the response times.

Figure 6.7 shows the measurement results for transaction $T2$ on the *Numbers* relations. The results are comparable to those discussed above, but the optimal number of processors is larger in the respective cases. This conforms with intuition: the overhead is equal, but the amount of data processing to be performed is larger due to the greater cardinalities of the involved relations. Note that the response times for 1 fragment are obscured in a negative sense by the system characteristics and limited resources of PRISMA/DB; clearly, a super-linear speedup is not reasonable in a well-parameterized system.

It is clear from these figures that response times of both unmodified (no constraint) and modified transactions can be reduced through the use of parallelism. Constraint enforcement profits most from parallelism in situations with constraints of 'average complexity': for simple constraints, the enforcement work is too small to profit from parallelism, and for complex constraints, the control overhead is too high because of the centralized transaction management.

The effect of main-memory data storage

In disk-based database systems, transaction response times are often dominated by disk I/O times. Clearly, main-memory systems can have an advantage here. The advantages for constraint enforcement of a main-memory system like PRISMA/DB are not as great as they may appear, however, because of two reasons. In the first place, the absence of stable main-memory causes the system to have partly disk-

6.4 Performance evaluation

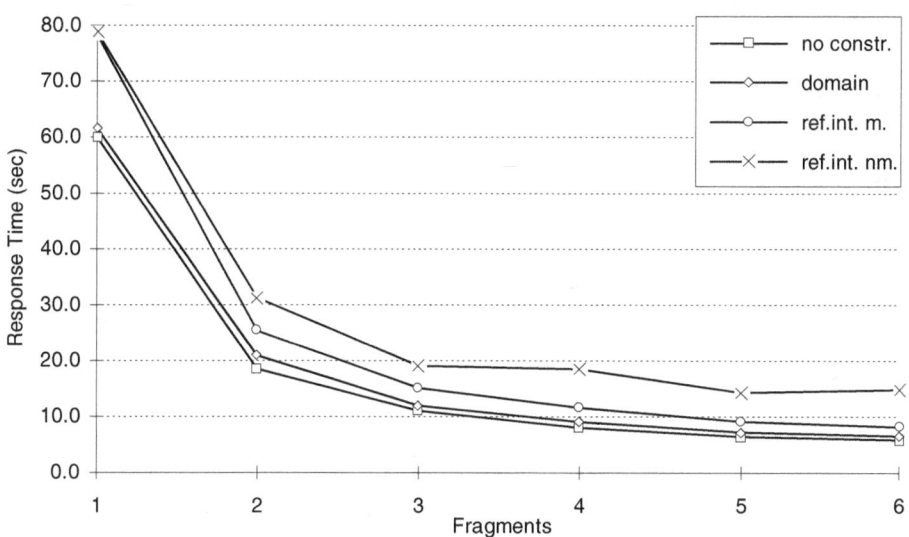

Figure 6.7: Speedup Numbers benchmark

based characteristics for update transactions, thereby reducing the advantage in transaction processing. In the second place, in disk-based systems the differential sets of update transactions are usually kept in buffers in main-memory, and do not have to be read from disk for constraint enforcement. So, constraint enforcement on differential sets only, like enforcement of domain constraints, does not profit from main-memory characteristics. In the case of multi-relation constraints, like referential integrity constraints, however, a main-memory system can profit from the fact that the relations not updated by the user transaction are already in main-memory.

Following the analysis above, only measurements with referential integrity constraint $I2$ are interesting here. Further, only the situation with the compatible fragmentation is taken into consideration, because the other situation would give the same differences between the main-memory and disk-based approach (since the same relation W3 has to be loaded from disk by the same number of data managers in both cases). The Wisconsin relation type is chosen for the experiments.

Since PRISMA/DB is a main-memory system, a disk-based system has to be simulated. This is accomplished here by removing the key relation ($W3$) of the referential integrity constraint from main-memory before transaction execution, such that it has to be loaded upon use.

The measurement results of the main-memory and disk-based situation are shown in Figure 6.8. As may be expected, the transaction execution time profits from the main-memory characteristics of PRISMA/DB. The advantage gets smaller, however, when the degree of parallelism increases. This is easily explained by the fact that the relatively expensive disk I/O is distributed over several parallel

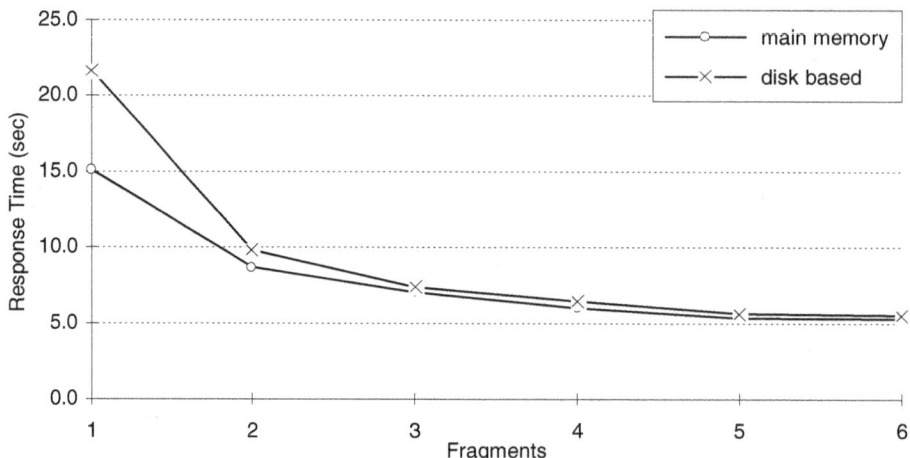

Figure 6.8: Main-memory versus disk-based data storage

processes.

6.4.4 Discussion of the results

This subsection presents some further discussion and analysis of some of the evaluation results presented above.

The effect of control overhead

The measurements presented above clearly show that parallelism is a good way to reduce response times of transactions including integrity control. They also show, however, that there is an optimum for the degree of parallelism in a certain situation. This is caused by the fact that more parallelism in the data management layer of PRISMA/DB implies more control overhead for the transaction manager process, which will always become a bottleneck at some point.

In a simple situation, the transaction manager overhead increases linearly with the number of processes to be controlled. The execution time per data manager decreases hyperbolically with the number of processes. The response time of a transaction can thus be modeled by the following function of the number of fragments f:

$$t_r(f) = A + B \times f + C + \frac{D}{f}$$

Here, A is the time for (startup) overhead of the transaction manager; B is the time per data manager spent by the transaction manager in setting up the execution infrastructure and sending the command; C is the time for (startup) overhead of the data manager; D is the execution time at the data management layer of the

6.4 Performance evaluation

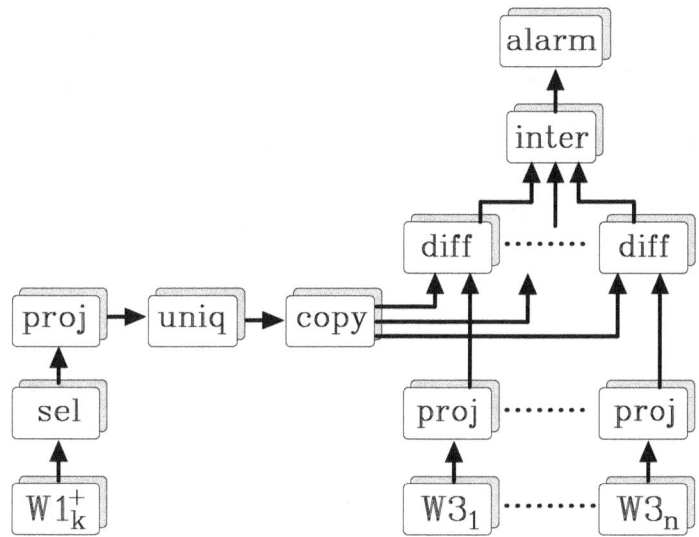

Figure 6.9: Execution infrastructure for referential integrity constraint

complete transaction (update plus logging/checkpointing) if performed by a single data manager. Clearly, this function has an optimum for a certain value of f.

In a more complex situation, the influence of the transaction manager overhead can get worse. For instance, in the case of a referential integrity constraint with incompatible fragmentation of key and foreign key relations, the overhead includes a quadratic factor because each foreign key fragment has to be matched against each key fragment. To illustrate this, the execution infrastructure for the enforcement of constraint $I2$ on one single fragment of relation $W1$ is shown in Figure 6.9.

The effect of a central data resource

As chosen in the benchmark design (see Figure 6.1), the data to be inserted into the relation to be modified is always stored at a single node. As such, this data forms a central resource, that may influence parallelism in the execution of the benchmark transactions in a negative sense.

To investigate this effect, the execution of transaction $T1$ with relation $W2$ non-fragmented is compared to the execution of $T1$ with $W2$ fragmented and allocated compatibly with respect to $W1$ (i.e. all insert operations can be performed completely locally on a node). Because this change only influences the execution of the insert operation and not the execution of constraint enforcement, only the situation without constraints is taken into account.

The measurement results are shown in Figure 6.10. Clearly, the situation with compatible fragmentation is the fastest, because no inter-node data transport is

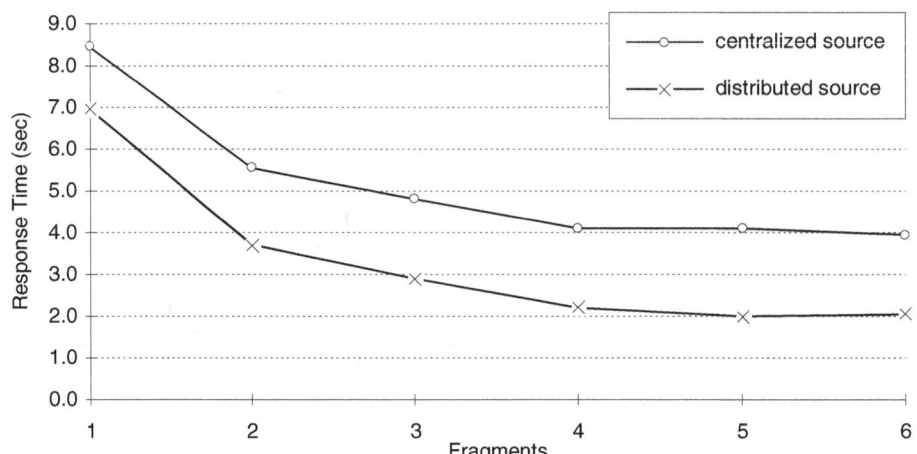

Figure 6.10: Centralized versus distributed source data

necessary. The speedup characteristics are about the same in both situations, however. This can be explained by the fact that the situation with compatible fragmentation does not include a central data resource, but does require more OFM processes to be managed by the transaction manager, which can become a bottleneck.

6.5 Conclusions

This chapter has discussed the implementation of an integrity control subsystem in a parallel main-memory database system. The subsystem is based on the transaction modification technique as described in the previous chapters.

As mentioned before, the prototype integrity control subsystem implemented in PRISMA/DB has a limited functionality. Extensions to a more complete functionality can easily be integrated in the current architecture. Due to the transaction modification approach, the major part of the extensions will apply to the constraint compiler component only. Changes to the data dictionary depend on the required functionality of the constraint analyzer component; they can be kept relatively small. Further, minor changes to the transaction manager will be necessary to handle multi-layer transaction modifiers. The other PRISMA/DB components can remain untouched.

The performance evaluation presented in this chapter leads to a number of important conclusions. In the first place, the overhead of constraint enforcement compared to the execution of transactions without integrity control is reasonable [Gref92b]. This demonstrates the feasability of the transaction modification approach in a real-world database system. In the second place, parallelism has proven to be an effective way to reduce response times of modified transactions. It has

6.5 Conclusions

to be remarked though, that the optimum level of parallelism is relatively small in the measurements. This is mainly due to the high costs of control overhead; ways to reduce these costs have been investigated in the PRISMA/DB context, but have not yet been applied to integrity control.

Chapter 7

Dynamic action scheduling

In the previous chapters, integrity control through transaction modification has been discussed in a parallel environment with fragmented relations. In this type of environment, modified transactions can easily get very complex. To make optimal use of parallelism to reduce transaction response times, advanced scheduling techniques are necessary for the execution of the statements in the transactions. These techniques must allow a high level of intra- and inter-transaction parallelism. This chapter discusses a graph-based approach to the design of these techniques.

The chapter starts with a short introduction to scheduling in parallel database systems and some basic notions. The next two sections discuss scheduling of actions within a transaction and scheduling of actions belonging to different transactions. The techniques are applied an in abstract system architecture. The results of testing these ideas on the PRISMA/DB platform are presented next. The chapter ends with a few conclusions.

7.1 Scheduling in parallel database systems

In general, (parallel) database systems are meant for high performance data processing . High performance can be seen as a combination of low response times and high throughput. To obtain these properties, a good scheduling of the actions on the database is essential. Two kinds of scheduling can be distinguished: scheduling of the actions in a single transaction, and scheduling of the actions of different transactions.

The scheduling of actions in a single transaction to obtain low response time is called *transaction optimization* here. The technique is based on dependencies between the various actions in a transaction, and takes the dynamic execution characteristics of the transaction into account, using availability of resources and feedback from the execution layer of the system. As such, the scheduling technique is based on a dynamic model of a multi-user machine, and can be seen as the complement of traditional query optimization techniques, which transform actions in a transaction based on a static model of a single-user machine. Transaction optimization is especially important in an environment with complex transactions, either user-defined or system-generated. The latter case occurs in distributed systems with fragmented relations [Ceri84] and in systems performing integrity constraint enforcement through transaction modification.

The scheduling of actions of different transactions is traditionally called *concurrency control*. Usually, this technique tries to optimize throughput under the condition that concurrently executing transactions are 'unaware' of each other. Concurrency control can be based on dependencies between the involved actions of concurrent transactions, much like the dependencies mentioned above. Transaction optimization and concurrency control can be integrated into one single mechanism that operates on a global graph representing the dependencies between actions. This leads to a system architecture with fully centralized action scheduling. Decentralizing this task can now be modeled conceptually by splitting up the graph into several subgraphs, operated upon by concurrency control and decentralized transaction management processes.

The approach to action scheduling presented in this chapter has a number of characteristics that distinguish it from the work in other research efforts [Gref91c, Gref92a]. In the first place, the approach makes use of high level dependencies between actions; other approaches, like [Eich88], use lower level dependencies representing the actual flow of data in a system. The use of high level dependencies enables simple scheduling mechanisms, causing little overhead and allowing for distribution of and parallelism in the scheduling task. In the second place, the scheduling technique presented in this chapter performs the dependency analysis of transactions dynamically at transaction execution time, thereby allowing parallelism between transaction control and transaction execution. Compiler-based techniques (e.g. [Hart88, Boral90]) perform the analysis at transaction definition time. The approach described here requires no compilation of transactions and is therefore suited for ad hoc transactions as well. Finally, the approach described in this chapter has been implemented and tested in a real-world parallel database machine. Other conceptually oriented approaches lack this practical test, whereas most database machine projects pay little attention to scheduling of concurrent multi-action transactions.

7.2 Basic notions

In Chapter 2 the extended relational algebra has been introduced to describe the actions in a transaction. Here, we are not interested in the complete semantics of algebra constructs, but merely in the characteristics of actions that are relevant to their execution order. Therefore, an abstraction is made from the extended relational algebra constructs to obtain a small set of elementary action types to specify transactions. The various action types are discussed below; the notation

begin	$begin$
assignment	$x = op(y_1, \ldots, y_n)$
output	$?x$
update	$upd(x, y)$
commit	$commit$

Table 7.1: Elementary action types

is shown in Table 7.1.

begin The begin action indicates the start of a transaction; it only has an action grouping purpose.

assignment The assigment action assigns the result of an extended relational operation to a new and implicitly defined relational variable; the semantics of the operation (join, union etc.) are of no significance in this context.

output The output action delivers the contents of a relational variable to the user of the database system; how the output is produced is of no significance; the output action models side effects of a transaction.

update The update action changes the current state of the database; the first parameter is the variable (relation) to be changed, the second parameter is a variable used as source for the change; the type of update (insert, delete or modify) is of no significance.

commit The commit action indicates the end of a transaction; it is only used for grouping purposes.

In this abstraction of the relational algebra, the stress is on operands, not on specific relational algebra operators. Note that this set of actions can be used to model more complex actions, such as actions with nested operations. Other choices of elementary action types are possible; the choice is not of great importance for the techniques presented in this chapter, however.

7.3 Order dependency and transaction optimization

This section discusses the dependencies between actions in a transaction with respect to their execution order, and the use of these dependencies in a scheduling algorithm that aims at an optimized execution of the transaction. The dependency concept and the various types of dependency are introduced first. The dependencies within a transaction can be represented by means of an order dependency graph. This graph is then used as the basis for the action scheduling algorithm.

7.3.1 Order dependency concept

Actions in a transaction can be dependent with regard to the order in which they have to be executed; this kind of dependency is called *order dependency*. The definition below defines the conditions under which two arbitrary actions in a single transaction are order dependent.

Definition 7.1 Given are the following two transactions T and T':

$$T = (a_1; \ldots; a_i; \ldots; a_j; \ldots; a_n)$$
$$T' = (a_1; \ldots; a_j; \ldots; a_i; \ldots; a_n)$$

Transaction T' is obtained from transaction T by interchanging actions a_i and a_j. Now action a_j has an *order dependency* with respect to action a_i in transaction T if a_i precedes a_j in the specification of T and at least one of the following holds:

- transactions T and T' model different transformations of at least one database state D:

$$(\exists D)(T(D) \neq T'(D))$$

- the side effects of the execution of T and T' on some database state D are different; in particular, T and T' either produce different output, or produce the same output in a different order.

The fact that a_j is order dependent on a_i is denoted as $od(a_j, a_i)$. □

This definition gives a conceptual view on order dependency, which is hard to use in a practical situation; therefore, a more operational approach is developed in the sequel of this chapter.

If the order dependency relations between actions of a transaction are to be analyzed, a minimal set of relations is preferable. Therefore, the definition of *direct order dependency* is presented below; this definition is a restriction of the one above.

Definition 7.2 Given is transaction T:

$$T = (a_1; \ldots; a_i; \ldots; a_j; \ldots; a_n)$$

Now action a_j has a *direct order dependency* with respect to action a_i if $od(a_j, a_i)$ and no action a_k exists with $i < k < j$ such that $od(a_j, a_k)$ and $od(a_k, a_i)$. The fact that a_j is direct order dependent on a_i is denoted as $dod(a_j, a_i)$. So, we have the following:

$$dod(a_j, a_i) \Leftrightarrow (od(a_j, a_i) \wedge \neg(\exists a_k)(i < k < j \wedge od(a_j, a_k) \wedge od(a_k, a_i)))$$

□

7.3 Order dependency and transaction optimization

	$begin$	$v = op$ $(w_1, \ldots, w_m)$	$?v$	$upd(v, w)$	$commit$
$begin$	-	-	-	-	-
$x = op(y_1, \ldots, y_n)$	$true$	$y_i \equiv v$	$false$	$y_i \equiv v$	-
$?x$	$true$	$x \equiv v$	$true$	$x \equiv v$	-
$upd(x, y)$	$true$	$x \equiv w_i$ $\vee y \equiv v$	$x \equiv v$	$x \equiv v \vee x \equiv w$ $\vee y \equiv v$	-
$commit$	$true$	$true$	$true$	$true$	-

Table 7.2: Order dependency matrix

7.3.2 Order dependency types

In a transaction consisting of a sequence of elementary actions as presented above, order dependency between two actions a_j and a_i exists in the following cases:

- If action a_i defines a variable that is used as an operand by action a_j, the actions cannot be interchanged without changing the effect of the transaction; this kind of order dependency is called *definition dependency*. Definition dependency models the *dataflow* between operations and is therefore also called *dataflow dependency*.

- If action a_i updates (modifies) a variable that is used as an operand by action a_j, interchanging the actions causes a_j to 'see' a wrong value of the variable; this kind of dependency is called *update dependency* or *value dependency*.

- If both actions a_i and a_j have side effects (produce output), interchanging the actions changes the order of the side effects; this kind of dependency is called *output dependency*.

Every possible ordered pair of elementary action types as defined above can be analyzed to obtain the conditions under which the two actions have an order dependency relation. These conditions are listed in Table 7.2[1]. A table of this kind can be used as a simple decision matrix to decide about the order dependency between two given actions.

7.3.3 Order dependency graph

The set of all direct order dependencies between the actions in a transaction can be described by means of a graph. The technique to represent sets of dependencies by means of a graph is commonly used in compiler construction, both for

---

[1] In this table the symbol $\equiv$ is to be interpreted as 'is the same variable as'. The actions in the rows are possibly dependent on the actions in the columns.

```
ROUTINE O-graph;
BEGIN
    V := {v_{a_1}, ..., v_{a_n}};
    E := ∅;
    FOR i FROM 2 TO n
    DO
        FOR j FROM i - 1 DOWNTO 1
        DO
            IF dod (action(v_{a_i}), action(v_{a_j}))
            THEN E := E ∪ {⟨v_{a_i}, v_{a_j}⟩}
            FI
        OD
    OD
END
```

Figure 7.1: O-graph construction algorithm

general-purpose programming languages [Aho78] and for database programming languages [Hart88]. The definition below describes a graph representing direct order dependencies.

Definition 7.3 Given is a transaction $T = (a_1, \ldots, a_n)$. A *Direct Order Dependency Graph* or O-graph of transaction T is a directed graph $G = \langle V, E \rangle$. The set of vertices V corresponds with the actions a_i in T: $V = \{v_1, \ldots, v_n\}$ and vertex v_i is labeled with action a_i. The set of edges E corresponds with the direct order dependencies that exist between the actions a_i of T: $E = \{\langle v_1, v_2 \rangle \in V \times V \mid dod(action(v_1), action(v_2))\}$. □

Constructing the O-graph from a given transaction $T = (a_1, \ldots, a_n)$ is rather straightforward; Figure 7.1 shows an algorithm in pseudo-code. Figure 7.2 shows an example transaction and the corresponding O-graph.

7.3.4 Dynamic transaction optimization

Dynamic transaction optimization is the technique of scheduling the execution of individual actions within a transaction such that the overall transaction execution response time is reduced, under the condition that the semantics of the transaction are not affected. Reduction of response time is obtained by scheduling the execution of the actions in a transaction as early as possible. Scheduling the sequential execution of actions implies monitoring the execution of the actions in a transaction and acting on the occurring events; transaction optimization is thus a dynamic process.

7.3 Order dependency and transaction optimization

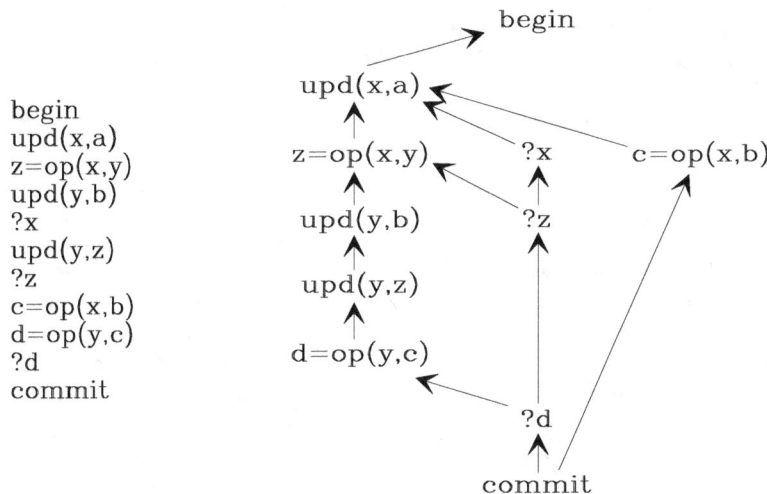

Figure 7.2: Example transaction and O-graph

Transaction optimization can be seen as a form of global query optimization [Satoh85]. As such, it is complementary to the usual query optimization techniques dealing with expression rewriting, common subexpression elimination and such (see for example [Ceri84]). Transaction optimization does *not* change the actions in a transaction in any way, but merely schedules their execution.

The issues of scheduling actions and maintaining transaction semantics are discussed in detail below.

Scheduling

Scheduling the actions of a transaction to reduce transaction response times implies deviating from a fully sequential execution of the actions in the order specified by the transaction, while taking order dependencies between the actions into account. There are a number of practical situations in which such a deviation is beneficial; these are discussed in detail below.

In case of *parallel processing* possibilities, several actions can be scheduled to be executed in parallel. Take the following simple example transaction:

begin
upd(a, b);
upd(c, d);
commit

The two update actions are fully independent, so they can be executed in parallel, thereby shortening the transaction execution time.

In case of *unavailable resources* necessary for some actions, the execution order of actions can be changed. Resources are mostly database data (relations or fragments of relations). This case is illustrated by the following example transaction:

begin
upd(a, c);
?c;
commit

If some other transaction holds a lock on resource a at the start of the execution of the transaction above, the first action cannot be executed, but the second can; therefore, changing the execution order will reduce transaction execution time.

In case of actions that may cause a *transaction abort*, these actions can be scheduled earlier to avoid unnecessary work in case of an abort. This situation occurs in a system that enforces integrity constraints through transaction modification: constraints are translated into extended relational algebra constructs that can contain an *alarm* operator triggering a transaction abort if some condition holds on their operands. Take the following example transaction:

begin
upd(a, b);
upd(c, d);
alarm(op(a));
commit

This transaction performs two updates on relations a and c. At the end of the transaction, an integrity constraint is evaluated over relation a. If this constraint is violated, the transaction will be aborted and the update on c has been superfluous work. Scheduling the *alarm* before the second update will avoid this situation.

Semantics

Scheduling the actions of a transaction may not affect the semantics of the transaction. This implies that it must be guaranteed that the optimized execution of a transaction has exactly the same effect as the fully serialized execution of the transaction, both in terms of database transition and side effects. As may be clear to the reader, the concept of order dependency is used for this purpose: actions in a transaction may not be scheduled such that two actions having an order dependency are executed in an other way than sequentially[2] in the order as indicated by the transaction.

As shown before, the dependencies in a transaction can be represented conveniently using an O-graph. Therefore, this graph will be used as the basis for the transaction optimization algorithms. This is discussed in the section below.

---

[2] As described later in this chapter (Section 7.7.2), some form of parallelism can be allowed between two actions that have a dataflow dependency. This is an operational aspect, however, that is not of importance on the conceptual level discussed here.

7.3.5 Scheduling using the O-graph

Transaction optimization is performed by scheduling algorithms operating on an O-graph. This requires the following elementary operations:

Adding a new vertex to the graph. A new vertex is added to the graph when a new action in the transaction is submitted to the system. Adding a new vertex implies adding all edges that originate from this vertex. This can be done using the decision matrix depicted in Table 7.2.

Updating the resource administration. When a new resource becomes available, the resource administration of the transaction is updated. This administration is used to check whether all resources for an action are available.

Submitting an action for execution. The action associated with a certain vertex is submitted for execution when the action is order executable (see below), and all necessary resources are available to the transaction.

Removing a vertex from the graph. A vertex is removed from the graph when the execution of the action associated with this vertex has been completed. Removing a vertex implies removing all edges ending in this vertex.

The description of the third operation above uses the concept *order executable* to indicate whether an action can be executed with respect to its order dependencies. This concept is defined formally as follows:

Definition 7.4 An action a_i of a transaction T is *order executable* if the vertex that corresponds with a_i has no outgoing edges in the O-graph $G = \langle V, E \rangle$. So:

$$oexec(a_i) \Leftrightarrow \{\langle v_1, v_2 \rangle \in E \mid action(v_1) = a_i\} = \emptyset$$

□

The process of action scheduling using the above operations on the O-graph is depicted in the flow diagram in Figure 7.3. The execution of the operations is of an event-driven nature.

7.4 Resource dependency and concurrency control

This section discusses the dependencies between actions belonging to different transactions that make use of the same resources (data). These *resource dependencies* can be represented by means of a graph. This graph can then be used for the scheduling of the actions to obtain a concurrency control protocol.

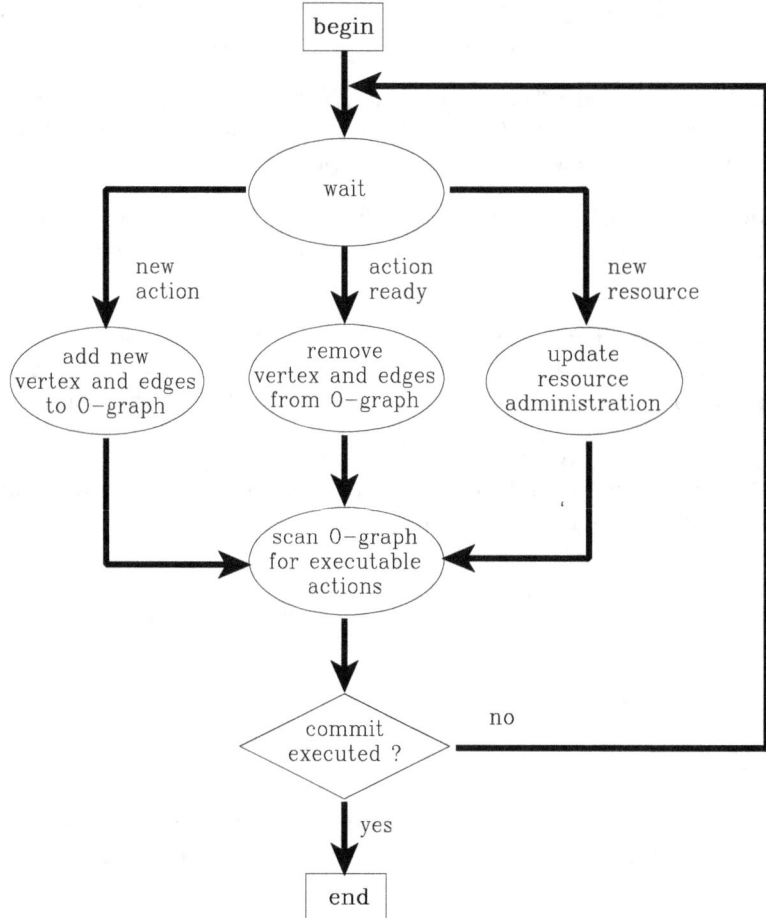

Figure 7.3: Action scheduling process using O-graph

7.4 Resource dependency and concurrency control

7.4.1 Resource dependency concept

Below the definitions of resource dependency and direct resource dependency are given; note that the definitions are analogous to the definitions of order dependency and direct order dependency as given before.

In short, two actions are *resource dependent* if an execution of these actions other than sequential and in the order in which the transactions to which they belong have started may violate the serializability property of the transactions the actions belong to.

Definition 7.5 Given are two transactions T_1 and T_2:

$$T_1 = (a_1; \ldots; a_i; \ldots; a_m)$$
$$T_2 = (b_1; \ldots; b_j; \ldots; b_n)$$

Now action a_i is said to have a *resource dependency* with respect to action b_j if the execution of a_i in transaction T_1 requires resources that are obtained or will be obtained by transaction T_2 and that cannot be released before the execution of action b_j has been completed. The fact that a_i is resource dependent on b_j is denoted as $rd(a_i, b_j)$. □

Definition 7.6 Given are two transactions T_1 and T_2 as shown above, and a set of transactions $\mathcal{T}$ being concurrently executed by the system. Now action a_i has a *direct resource dependency* with respect to action b_j, if $rd(a_i, b_j)$ and no action c_k exists in any transaction $T_x \in \mathcal{T}$, such that both $rd(a_i, c_k)$ and $rd(c_k, b_j)$. The fact that a_i is direct resource dependent on b_j is denoted as $drd(a_i, b_j)$. So we have:

$$drd(a_i, b_j) \Leftrightarrow (rd(a_i, b_j) \land (\forall T_x \in \mathcal{T})(\neg(\exists c_k \in T_x)(rd(a_i, c_k) \land rd(c_k, b_j))))$$

□

As will be clear, resource dependencies between actions in two transactions can be analyzed only during the execution of these transactions, whereas order dependencies as discussed before can be analyzed statically.

7.4.2 Resource dependency graph

The set of all direct resource dependencies between the actions of transactions being executed by the system at a given time can be described by means of a graph, called R-graph. This graph is similar to the Wait-For-Graph (WFG) used commonly in concurrency control [Date83, Ceri84]. The R-graph as defined below contains more information than a WFG, however, since its nodes are not transactions, but actions of the transactions. In other words, the R-graph represents dependencies at a more detailed level than the WFG.

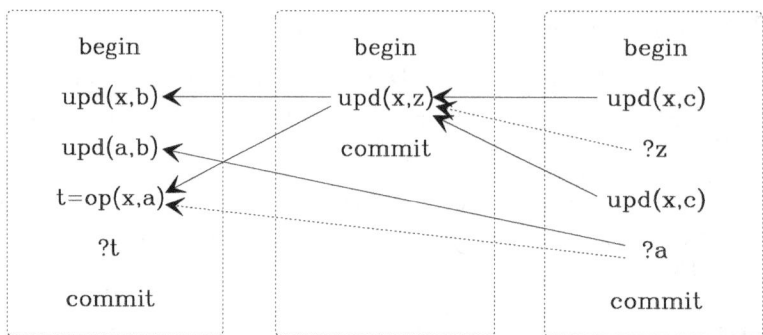

Figure 7.4: Example transactions and R-graph

Definition 7.7 Given is a set of transactions $\mathcal{T} = \{T_1, \ldots, T_m\}$ being executed by the system at a given time, with $T_i = (a_1^i, \ldots, a_{n_i}^i)$. A *Direct Resource Dependency Graph* or R-graph of a transaction set $\mathcal{T}$ is a directed graph $G = \langle V, E \rangle$. The set of vertices V corresponds with the actions a_j^i of the transactions in $\mathcal{T}$: $V = \{v_j^i \mid 1 \leq i \leq m \wedge 1 \leq j \leq n_i\}$ and vertex v_j^i is labeled with action a_j^i. The set of edges E corresponds with the direct resource dependencies that exist between the actions a_j^i of $\mathcal{T}$: $E = \{\langle v_1, v_2 \rangle \in V \times V \mid drd(action(v_1), action(v_2))\}$. □

The use of an R-graph is independent of the locking scheme used. Figure 7.4 shows an example of an R-graph with three concurrently executing transactions. The resource dependencies associated with the solid edges are based on a two-phase locking protocol with shared and exclusive locks [Date83]. If exclusive locks are used only, the dependencies associated with the dotted edges are added to the graph.

7.4.3 Concurrency control

As shown above, resource dependencies between actions of various transactions can be represented conveniently by means of an R-graph. Therefore, this graph can easily be used for concurrency control purposes. Concurrency control is then performed by manipulating the R-graph and keeping a resource administration. Manipulating the R-graph consists of the following basic operations:

Adding a new vertex to the graph. A new vertex is added to the R-graph when a new action in a transaction is to be executed. This can be done at two different moments: when the action is submitted to the system, and when the transaction is ready to actually execute the action. These situations can be described as 'greedy locking', respectively 'lazy locking'. Inserting a new vertex implies inserting all edges originating from this vertex.

7.5 Integrating both worlds

Submitting an action for execution. After a vertex has been added to or removed from the R-graph, the graph is scanned for actions that can be submitted for execution, i.e. actions that are *resource executable*. When an action is submitted for execution, the resource administration may need to be updated (resources may have become unavailable).

Removing a vertex from the graph. A vertex associated with a certain action is removed from the R-graph when the execution of that action has been completed. Removing a vertex implies removing all edges ending in this vertex. Further, the resource administration may need to be updated (resources may have become available).

The concept resource executable as used above is defined as follows:

Definition 7.8 An action a_i of a transaction T is *resource executable* if the vertex that corresponds with a_i has no outgoing edges in the R-graph $G = \langle V, E \rangle$. So:

$$rexec(a_i) \Leftrightarrow \{\langle v_1, v_2 \rangle \in E \mid action(v_1) = a_i\} = \emptyset$$

□

7.5 Integrating both worlds

In the previous two sections the notions of order and resource dependency and their graph representations were discussed; as mentioned before, the notions and their graph representations are much alike. Therefore, both types of dependencies are integrated into one global graph representation, that describes all dependencies between actions being handled by the system.

7.5.1 Global dependency graph

A G-graph is a graph that represents all dependencies between the actions in a set of transactions being executed. It is defined as follows:

Definition 7.9 Given is a set of transactions $T = \{T_1, \ldots, T_m\}$ being executed by the system at a given time, with $T_i = (a_1^i, \ldots, a_{n_i}^i)$. A *Global Direct Dependency Graph* or G-graph of transaction set T is a directed graph $G = \langle V, E \rangle$. The set of vertices V corresponds with the actions a_j^i of the transactions in T: $V = \{v_j^i \mid 1 \leq i \leq m \wedge 1 \leq j \leq n_i\}$ and vertex v_j^i is labeled with action a_j^i. The set of edges corresponds with the direct dependencies that exist between the actions of T: $E = \{\langle v_1, v_2 \rangle \mid dd(action(v_1), action(v_2))\}$. The *direct dependency* relation between two actions is defined as follows:

$$dd(v_1, v_2) \Leftrightarrow dod(v_1, v_2) \vee$$
$$(drd(v_1, v_2) \wedge \neg(\exists v_3 \in V \mid od(v_1, v_3) \wedge rd(v_3, v_2)))$$

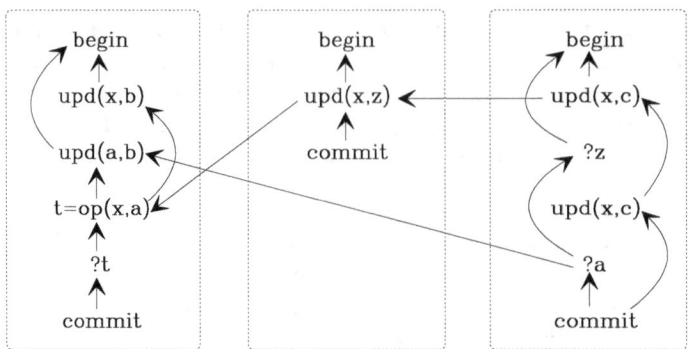

Figure 7.5: Example transactions and G-graph

Informally, a G-graph is constructed by merging the R-graph and all the O-graphs of the transactions being executed, and removing all superfluous resource dependencies. A resource dependency is superfluous here, if it indicates that an action a_1 must wait for a resource while an action a_2 that is surely executed before a_1 is waiting for the same resource. Figure 7.5 shows an example of a G-graph with three transactions, based on the R-graph shown before. In this graph, edges within the boundaries of the transactions represent direct order dependencies; edges crossing transaction boundaries represent direct resource dependencies.

7.5.2 Global scheduling

The G-graph represents all dependencies between actions in transactions being executed by the system. Global scheduling of the execution of actions can therefore be based on the G-graph. The scheduling is analogous to the scheduling based on an O-graph or R-graph: actions submitted to the system are added to the graph, executable actions are submitted to the action execution layer, and completed actions are removed from the graph. The following definition states when an action is (globally) executable.

Definition 7.10 An action a_i of a transaction T is *globally executable* if the vertex that corresponds with a_i has no outgoing edges in the G-graph $G = \langle V, E \rangle$. So:

$$gexec(a_i) \Leftrightarrow \{\langle v_1, v_2 \rangle \in E \mid action(v_1) = a_i\} = \emptyset$$

Global scheduling based on the G-graph implements both dynamic transaction optimization and concurrency control within one conceptual mechanism.

7.6 Architectural issues

The previous sections of this chapter have discussed algorithms for dynamic action scheduling. In this section the architecture of an action scheduler component and its integration into a database management system are discussed at a conceptual level. These ideas are used in Section 7.7 in a real world DBMS.

7.6.1 Action scheduler architecture

An architecture for the Action Scheduler is shown in Figure 7.6. The following components can be identified:

Graph Processor The graph processor forms the heart of the action scheduler; this module maintains the dependency graph using the algorithms described before in this thesis. The processor receives new actions to be added to the graph when new tasks are submitted to the action scheduler. Actions that are executable are sent for execution to the execution control. When the execution of an action has been completed, the action is removed from the graph; its resources are handed back to the resource control module.

Action Analyzer The Action Analyzer module analyzes incoming actions prior to sending them to the graph processor. The analysis detects the resources necessary for the execution of the actions; resource requests are sent to the resource control module.

Resource Controller The Resource Control module keeps an administration of the resources needed for transaction execution. If it does not manage all resources in the system, it can take steps to acquire resources from other resource controllers. Available resources are sent to the graph processor.

Execution Controller The Execution Control module controls the execution of actions that were released by the graph processor. It monitors the execution, such that completion of actions can be notified to the graph processor.

Messages coming in at one of the three interfaces to the graph processor form the events that trigger the activities of this module.

7.6.2 Integrating the action scheduler into a DBMS

The action scheduler as discussed above can be integrated into a complete DBMS architecture. The most simple approach is to have one central action scheduler in the system. In the case of a distributed system, it can be advantageous to split up the scheduler into a number of schedulers that each perform part of the scheduling task.

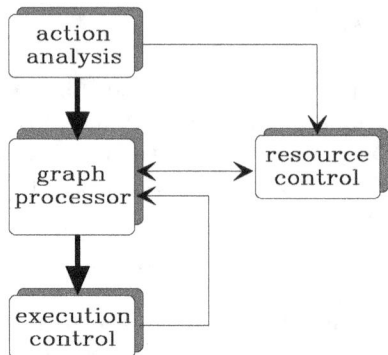

Figure 7.6: Action Scheduler architecture

Centralized action scheduling

An action scheduler managing the entire G-graph of a system can be used as a centralized transaction management layer of the DBMS, controlling both transaction execution and concurrency between transactions. This situation is depicted in Figure 7.7.

The action scheduler forms the interface between the action preprocessing layer and the action execution layer of the system. It accepts transaction specifications from the query optimizer that are in a ready to use and optimized form, and submits actions to be executed to the execution layer of the system.

Note that pipelining is possible in this process: the transactions can be handed in several pieces to the action scheduler, and scheduling can start immediately when a piece is available. In this way, parallelism can be obtained between action preprocessing, action scheduling and action execution. Parallelism between the various query processing layers of a DBMS can result in an improved overall performance (see for example [Li88]).

Distributed action scheduling

In a distributed (parallel) DBMS it can be advantageous to distribute the action scheduling tasks to avoid that the scheduler becomes a performance bottleneck. Distributing the scheduling tasks can easily be done by partitioning the dependency graph and assigning a private graph processor to each partition. The graph processors can then be allocated on different processors of the hardware architecture.

A natural form of distribution is obtained by partitioning the central G-graph into one central R-graph and an O-graph for each transaction being executed. This leads to a situation with distributed transaction management and centralized concurrency control. The corresponding system architecture is depicted in Figure 7.8.

7.6 Architectural issues

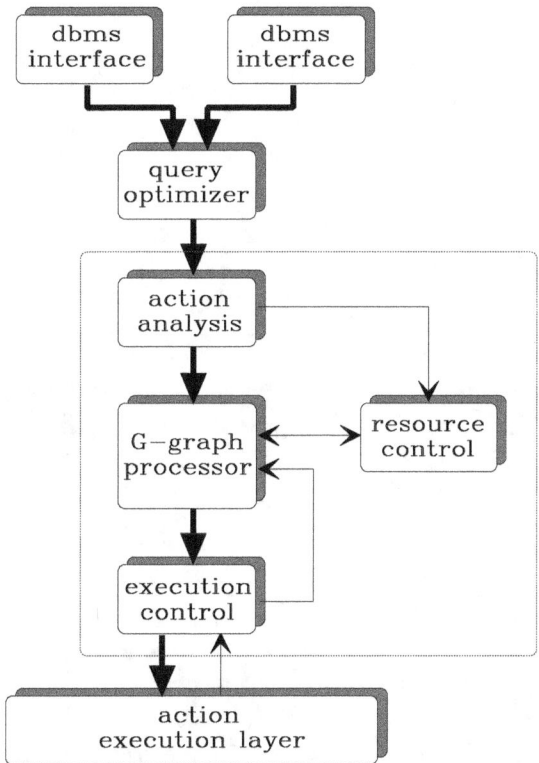

Figure 7.7: Action scheduling with centralized transaction management

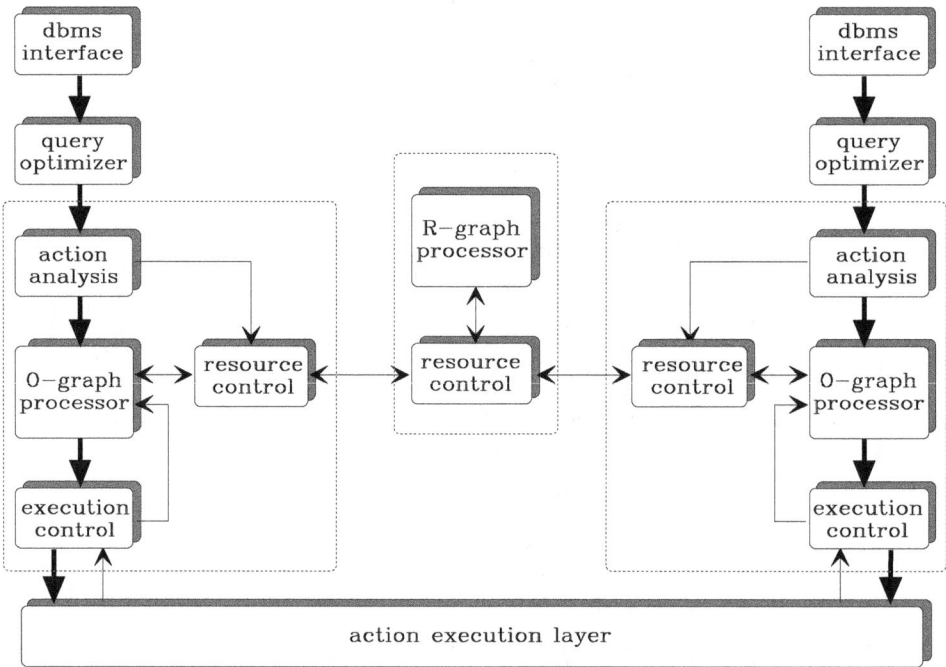

Figure 7.8: Action scheduling with distributed transaction management

7.7 Action scheduling in PRISMA/DB

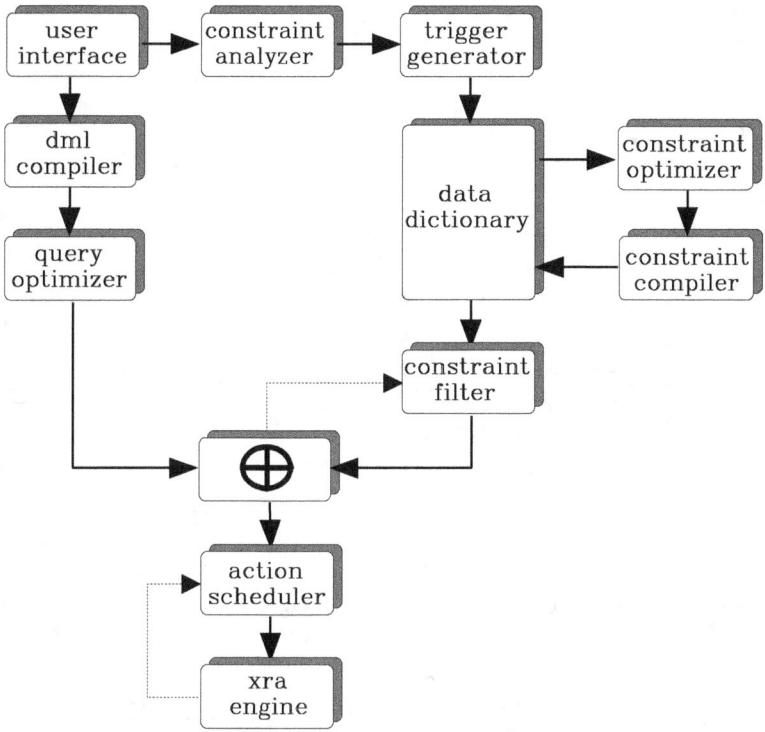

Figure 7.9: Action scheduling in abstract architecture

It is possible to distribute the central R-graph also to obtain a situation with both distributed transaction management and distributed concurrency control.

Integration in transaction modification

To demonstrate the compatibility of the techniques of this chapter with the integrity control ideas of Chapters 4 and 5, Figure 7.9 shows how an action scheduler component can be integrated into the abstract system architecture for transaction modification.

7.7 Action scheduling in PRISMA/DB

The action scheduler architectures as depicted in Figures 7.7 and 7.8 can be used in real world database systems. This section discusses the application of the ideas in the context of the PRISMA/DB parallel database management system. First, the necessary extension to the PRISMA/DB architecture is discussed. Next, measurements on the system are presented.

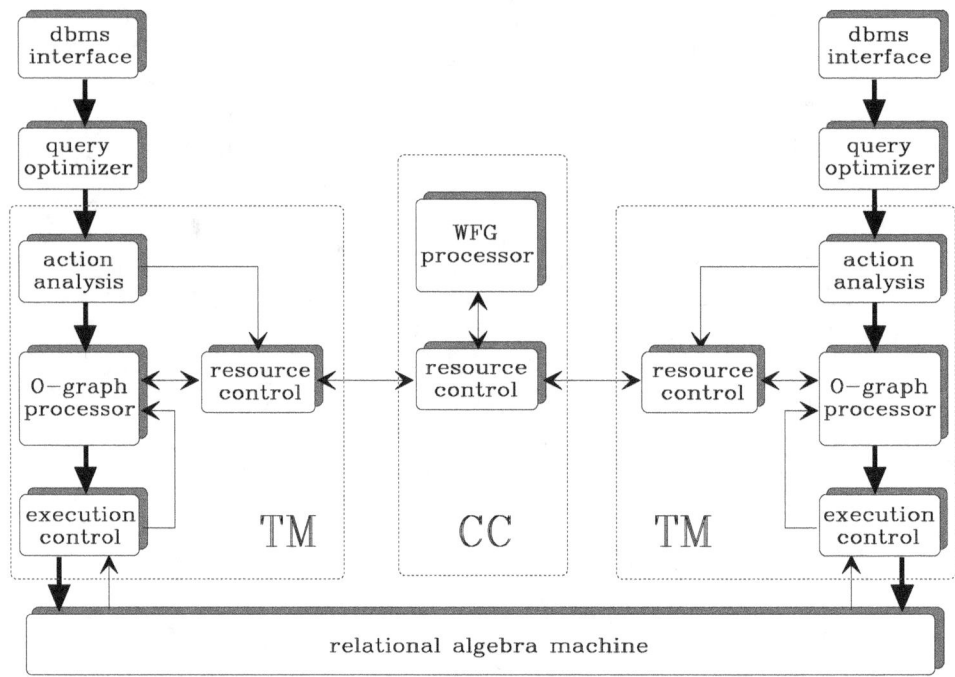

Figure 7.10: Action scheduling in PRISMA/DB

7.7.1 Architecture

As described in Chapter 6, PRISMA/DB is a parallel database management system with distributed transaction management and centralized concurrency control. Therefore, the action scheduling architecture shown in Figure 7.8 can easily be mapped to the PRISMA/DB architecture.

The O-graph processor is located in the transaction manager component (TM). It uses a variation on the graph processing algorithms presented in Section 7.3.5. The used protocols allow for parallelism between the action scheduler (TM) and action execution layers, such that part of the scheduling can take place when execution is on the way. Details of the graph processor implementation can be found in [Gref91c]. The R-graph processor is located in the concurrency control component (CC). In cooperation with the TM's, the CC employs a simple two-phase locking protocol with shared and exclusive locks [Date83, Ceri84]. Locks are always released at transaction commit. This implies that all edges in the R-graph end in a *commit* action, and that the R-graph is equivalent to a Wait-For-Graph. A centralized CC process is used because this simplifies the design of PRISMA/DB, and enables cheap deadlock prevention (since the entire R-graph is located on one node of the system).

7.7 Action scheduling in PRISMA/DB

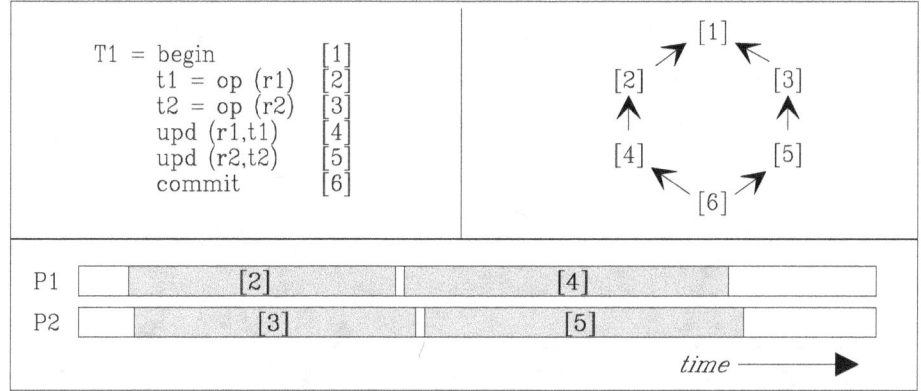

Figure 7.11: Scheduling of a transaction

7.7.2 Measurement results

This section presents the results of measurements to show the effectiveness of transaction optimization in the PRISMA/DB context. The goal of this section is to give the reader a general impression, not to present a complete performance analysis. Two situations are discussed:

- the situation in which parallel execution of actions is used to reduce the execution time of a transaction;
- the situation in which the order of execution of actions is changed because of unavailable resources.

Currently, PRISMA/DB does not make use of early abort scheduling, so this situation cannot be demonstrated here. Further details on the experiments can be found in [Gref91c].

Parallelism

Figure 7.11 shows the execution of a transaction $T1$. The upper left part of the figure shows the transaction in terms of the action types presented in Section 7.2. The O-graph of the transaction is depicted in the upper right part of the figure. The lower part of the figure shows how the execution of the actions takes place in time on the processors of the system. Each bar represents one processor executing actions of the transaction. The length of the bars represents the total execution time of the transaction, including control overhead at the beginning of the transaction and logging at the end of the transaction. Only the execution of the actions is shown in the bars; the scheduling of the transaction takes place on a different processor and is not shown.

Transaction $T1$ as shown in Figure 7.11 performs actions on two relations $r1$ and $r2$ allocated on two processors $P1$ and $P2$. The actions on $r1$ and those on $r2$

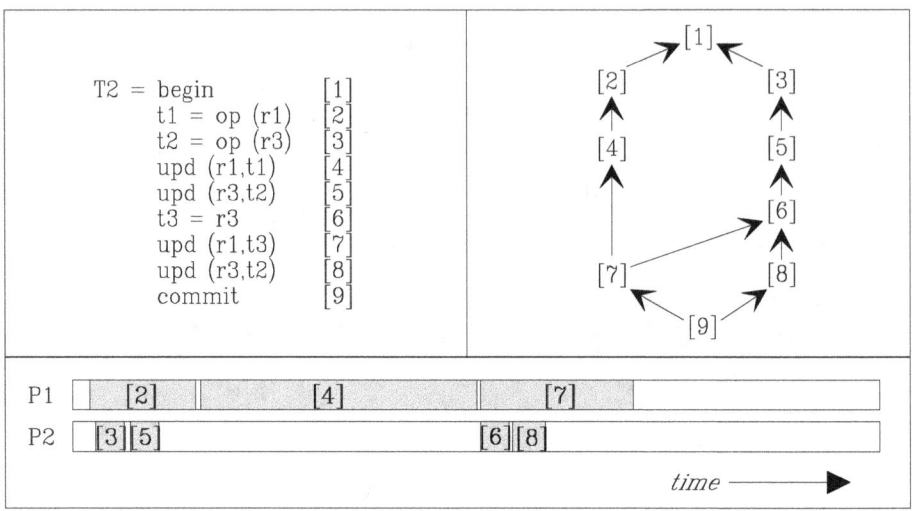

Figure 7.12: Scheduling of a transaction

are mutually independent, so they can be executed in parallel. Figure 7.11 shows that this is indeed the case. The gain in transaction execution time compared to a sequential execution of all actions is obvious.

Figure 7.12 shows the execution of a transaction in which order dependencies exist between the actions on the relations involved. Transaction $T2$ performs actions on two relations $r1$ and $r3$. Actions [2] through [5] can be scheduled in the same way as in the previous example. As shown in the O-graph, action [7] is order dependent on action [6]. Action [6], however, involves the transfer of data between processors $P1$ and $P2$ (modelled here as a 'remote' assignment), and cannot be executed before $P2$ is ready to receive data, i.e. has executed action [4]. The fact that PRISMA/DB exploits pipelining parallelism [Gref88, Wils89, Wils90] enables a parallel execution of actions [6] and [7]. This does not violate the theory of order dependency. Action [8] is order dependent on action [6], and has to wait therefore for the completion of this action. This example shows again, that the execution of actions is scheduled as early as possible, based on the dependencies between them.

Concurrency

In Figure 7.13 the execution of two concurrent transactions $T3$ and $T4$ is depicted [3]. Transaction $T3$ is started slightly earlier than $T4$ in this example. Therefore, $T3$ first obtains an exclusive lock on relation $r1$. Resource $r1$ is unavailable at the start of $T4$, so the actions of $T4$ on $r2$ are scheduled earlier than those on $r1$. The actions of $T4$ on $r1$ are executed as soon as $T3$ has released its locks. Note

---

[3] As will be clear, the graph in this figure is a G-graph, not an O-graph like in the preceding examples.

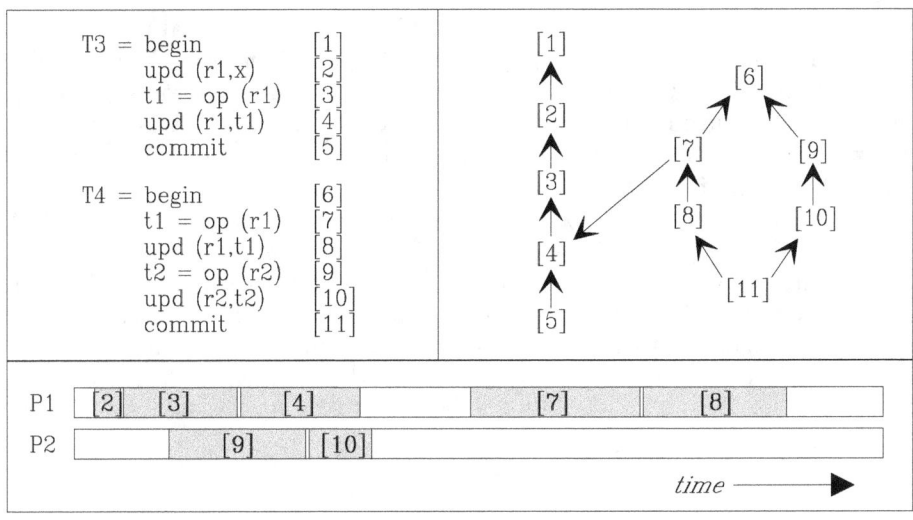

Figure 7.13: Scheduling of concurrent transactions

that $T3$ has to log its updates before it can release its locks on relation $r1$ [4]; this accounts for the time gap between the execution of actions [4] and [7].

7.8 Conclusions

This chapter describes a dynamic action scheduling technique that makes use of parallel action execution, resource availability information and early abort situations to improve both the response times of transactions and the throughput of a database system. This results in an improvement of the overall system performance. The scheduling technique will be most beneficial in multi-user systems with complex transactions, i.e. transactions consisting of many actions. These complex transactions may be defined by the user, or generated by the system. The latter case occurs in systems with fragmented relations, and in systems that enforce integrity constraints via the normal transaction mechanisms (e.g. constraint enforcement through transaction modification). Clearly, scheduling is beneficial in the type of system discussed in the preceding chapters, as it has both properties.

Dynamic action scheduling can easily be described using a graph-based approach. The scheduling algorithms can be cast easily into graph processor algorithms, used for a graph-based action scheduler. This scheduler can be integrated into an abstract DBMS architecture. To accomodate decentralized transaction management, the scheduler can be decentralized by partitioning the global graph it operates upon. The feasibility of the approach is demonstrated by the implemen-

---

[4]This is due to the fact that transactions in PRISMA/DB do not release any locks before end of transaction, and logging is considered an integral part of a transaction performing updates.

tation of a decentralized graph processor in the PRISMA parallel database system. Measurements performed on this system show the effectiveness of the scheduling technique, both for reducing transaction response times and for improving system throughput. The scheduling overhead in transaction response times is low, due to the simple scheduling algorithms, and to the fact that transaction scheduling and execution can be performed in parallel.

The technique can easily be extended and improved in a number of ways. First, the analysis of the dependencies between actions can be made more 'intelligent'. The algorithms may detect for example, that two insert operations on the same relation are not order dependent, because they use independent sources. Secondly, the scheduling algorithms can use resource information not only concerning data resources, but also concerning processing resources (i.e. processor load).

Chapter 8

Extending the ideas

The preceding chapters of this thesis have given a complete overview of the transaction modification approach to integrity control in parallel database systems. The concepts and techniques presented form a basic collection, however, that may need extension for specific application environments. This chapter presents ideas for extensions in different directions. The purpose of this chapter is to demonstrate the extensibility of the transaction modification approach. Therefore, the presented ideas form the starting point for a more complete design of the extensions and may be seen as an introduction to future research.

A rather obvious extension is the handling of extended integrity constraint types, i.e. constraint types that cannot be described in the $\mathcal{CL}$ language presented in Chapter 3. This topic is discussed in the first section of this chapter.

Integrity constraint enforcement can be seen as the automatic execution of actions that verify database states and transitions. This idea can be extended to other automatic actions to be performed by a database system, leading to the notion of active databases. Extending the transaction modification technique to the active database domain is discussed in Section 8.2.

The last section of this chapter discusses constraint enforcement protocol extensions to be used in environments with high performance requirements. The protocol extensions can be seen as complementary to the transaction modification techniques for constraint enforcement as discussed in the preceding chapters.

Note that other extensions of the techniques are possible too, for instance the use of transaction modification for nested transactions [Beeri88, Schön89]. The extensions presented in this chapter are to be considered as realistic examples, demonstrating how new ideas can be integrated into the transaction modification approach.

8.1 Extended constraint types

In Chapter 3, the constraint specification language $\mathcal{CL}$ has been introduced. Although it is a powerful language, suitable for a broad range of state and transition constraints, not all constraints can be expressed in it. Examples of constraint types that cannot be expressed in $\mathcal{CL}$ are constraints on partitions (groups) of relations and constraints with a recursive nature. This section discusses extensions of $\mathcal{CL}$ that enable these constraint types, and their translation to extended relational algebra.

8.1.1 Extended aggregate constraints

The aggregate constraint constructs introduced in Chapter 3 only allow aggregate functions over entire multi-sets. In many applications, however, a multi-set can be subdivided into a number of sub-multi-sets based on the value of a certain attribute, and an aggregate function must be evaluated over each of these sub-multi-sets. Below, language $\mathcal{CL}$ is extended to language $\mathcal{CL}_A$ to include these constructs.

Definition 8.1 The constraint specification language $\mathcal{CL}_A$ is obtained from language $\mathcal{CL}$ as defined in Chapter 3 (Definitions 3.2-3.5) by the following modifications:

- The alphabet of $\mathcal{CL}$ is extended with the set of multi-set function symbols $FM = \{\sigma\}$.

- The set of multi-set terms $\mathcal{T}_M$ is introduced; the following constructs are elements of this set:
  - A multi-set constant from the set M.
  - A multi-set function application $\sigma_\varphi T$, with $T \in \mathcal{T}_M$ and φ a condition on elements of M.

- The set of terms $\mathcal{T}$ is renamed into the set of value terms $\mathcal{T}_V$; this set is redefined as follows:
  - A value constant from the set C.
  - An attribute selection $x.i$, where $x \in V$ and i an integer constant from C.
  - An arithmetic function application $t_1 \vartheta t_2$, where $\vartheta \in FV$, and $t_1, t_2 \in \mathcal{T}_V$.
  - An aggregate function application $\Gamma(R, i)$, with $\Gamma \in FA$, $R \in \mathcal{T}_M$, and i an integer constant from C.
  - A counting function application $\Gamma(R)$, with $\Gamma \in FC$ and $R \in \mathcal{T}_M$.

- The set of atomic formulas $\mathcal{A}$ is redefined as follows:

8.1 Extended constraint types

- An arithmetic comparison $T_1 \vartheta T_2$, with $\vartheta \in PV$, and $T_1, T_2 \in \mathcal{T}_V$.
- A multi-set membership expression $x \in R$, where $x \in V$, and $R \in \mathcal{T}_M$.
- A tuple comparison $x = y$, where $x, y \in V$.

□

In short, this extension of $\mathcal{CL}$ extends the multi-set construct with selections on multi-sets. This construct is illustrated below:

Example 8.1 An example constraint using the constructs introduced above is the following: every brewery in the database should brew at least two beers; this constraint can be formulated in $\mathcal{CL}_A$ as:

$$(\forall x)(x \in beer \Rightarrow CNT(\sigma_{brewery=x.brewery} beer) \geq 2)$$

If used in an aborting integrity rule, this constraint can be translated to the following extended relational algebra construct:

$$alarm(\sigma_{\neg(\%2 \geq 2)}(groupby((brewery), CNT, name, beer)))$$

□

8.1.2 Recursive constraints

The $\mathcal{CL}$ constraint specification language does not include any facilities for the specification of recursive constraints. Many applications can, however, benefit from the possibility of recursion in constraint definitions. Examples are to be found in deductive databases, railroad network databases, part-of databases, all of which may contain recursive relationships between data. Here, we discuss a simple though powerful extension of the $\mathcal{CL}$ language: the transitive closure. The extension to the constraint specification language is discussed below. As the extended relational algebra, into which constraints are translated, also lacks recursion, this language is extended first. Finally, an example constraint and its translation are presented.

Relational algebra extension

To enable the specification of recursively defined multi-sets in the extended relational algebra, the transitive closure operator is added to the definition of the extended relational expressions as follows:

Definition 8.2 The *extended relational algebra expressions* are redefined as follows. Let E denote an extended relational algebra expression defined on schema $\mathcal{E}$. Then the following constructs are extended relational algebra expressions:

- The extended relational algebra expressions as defined in Chapter 2, Definition 2.9.

- The *transitive closure* expression $closure(\alpha, \varphi, E)$, where attribute list α with schema $\mathcal{E}$ specifies a projection on $\mathcal{E} \times \mathcal{E}$, and φ is a condition defined on elements of $\mathcal{E} \times \mathcal{E}$. The semantics of the *closure* operation is defined by the algorithm below:

$$new = E;$$
$$pow = E;$$
$$REPEAT$$
$$\quad old = new;$$
$$\quad pow = \pi_\alpha(E \bowtie_\varphi pow);$$
$$\quad new = new \cup pow;$$
$$UNTIL\ old = new$$

$\square$

Constraint specification language extension

Like the $\mathcal{CL}$ language was extended to the $\mathcal{CL}_A$ language by the introduction of the multi-set *select* function, the language is here extended again by adding a *transitive closure function*. The extension is defined as follows:

Definition 8.3 The constraint specification language $\mathcal{CL}_C$ is obtained from language $\mathcal{CL}_A$ (see Definition 8.1) by the following modifications:

- The set of multi-set function symbols is extended with the transitive closure function symbol Δ to $FM = \{\sigma, \Delta\}$.

- The set of multi-set terms T_M is extended with the transitive closure construct:
    - A multi-set function application $\Delta_\varphi^\alpha T$, with $T \in T_M$, φ a condition defined on $dom(T) \times dom(T)$, and α an attribute list defined on $dom(T) \times dom(T)$ with schema $dom(T)$.

$\square$

The semantics of the the Δ function are identical to the semantics of the *closure* operator from the extended relational algebra. The application of the transitive closure is illustrated by the example below.

Example 8.2 The *brewery* relation of the example beer database has a *part_of* attribute, indicating to what larger brewing company a brewery belongs (see Appendix B). Clearly, a brewery cannot be part of itself; this is easily expressed as a tuple constraint in $\mathcal{CL}$:

$$(\forall x)(x \in brewery \Rightarrow x.name \neq x.part\_of)$$

But if the larger brewing company can itself be part of an even larger company and so on, the constraint above is not strong enough. The following transitive closure constraint states that a brewery cannot be a part of itself, neither directly nor indirectly:

$$(\forall x)(x \in \Delta_{\%4=\%5}^{(\%1,\%2,\%3,\%8)} brewery \Rightarrow x.name \neq x.part\_of)$$

This constraint can be translated to the following extended relational algebra construct:

$$alarm(\sigma_{name=part\_of} closure((\%1,\%2,\%3,\%8),\%4=\%8, brewery))$$

□

8.2 Active databases

In many application areas, it is very useful if the database system can automatically respond to certain operations performed against the database. Integrity control is an important example: whenever incorrect updates are applied to the database, the integrity control system will automatically bring the database into a correct state. Other automatic actions are for example derived data maintenance and data protection or authorization control. Database systems that can automatically perform a number of such tasks are called *active database systems*. Recent systems with this functionality are POSTGRES [Stone86b, Stone88, Stone90a] and Starburst [Haas90, Lohm91, Widom91]. Both systems use a rule system as the basis for the active database functionality.

The transaction modification technique can also be used to equip a database management system with active database features. This is illustrated in this section by means of a number of examples.

8.2.1 Derived data maintenance

Derived data in a database can be defined by means of normal views, i.e. expressions on the base relations in the database. Views can be seen as virtual relations, i.e. relations that are not actually stored in the database, but derived from the base data whenever needed. Clearly, if a view is used frequently and the calculation of the view data is costly, this is not an efficient way to handle derived data. A solution is to work with materialized views, i.e. derived data that is stored in the database. This data has to be maintained automatically, however, upon changes to the relations on which the view is defined [Ceri91]. The maintenance of the materialized views can be realized by means of transaction modification. This is illustrated below.

Example 8.3 Suppose that the example beer database is often used to retrieve the names of Dutch beers. Then the following materialized view could be defined:

DEFINE MATERIALIZED VIEW *dutch_beers* AS
$\pi_{beer.name}(beer \bowtie_{beer.brewery=brewery.name} (\sigma_{country="NL"} brewery))$

This materialized view can be maintained by the following rules:

WHEN $INS(beer)$
IF NOT $false$
THEN $new =$
$\quad \pi_{beer.name}(beer^+ \bowtie_{beer.brewery=brewery.name} (\sigma_{country="NL"} brewery));$
$\quad insert(dutch\_beers, new)$

WHEN $DEL(beer)$
IF NOT $false$
THEN $delete(dutch\_beers, \pi_{name} beer^-)$

□

Note how the actions of the rules can make use of the differential sets of relation *beer* to avoid unnecessary work. The translation to triggered programs[1] to be used in transaction modification is trivial (just leave out the condition part of the rules).

The transaction modification technique can also be used to prevent user updates against materialized views. For this purpose, the actions of the maintenance rules for the derived views are declared to be non-triggering (see Section 4.6), and update prevention rules are defined.

Example 8.4 The materialized view defined in the preceding example can be protected against corrupting updates by means of the following rule:

WHEN $INS(dutch\_beers), DEL(dutch\_beers)$
IF NOT $false$
THEN $abort$

□

This update prevention mechanism can be seen as a simple form of access control. The materialized view maintenance and update prevention rules can easily be automatically derived by the system from the definition of the view.

---
[1]The term *triggered program* is used here for the combination of a trigger set and an extended relational algebra program. In Chapter 4, the term *integrity program* is used, but this term is hardly applicable here.

8.2.2 System logging

In a number of application domains, it is useful if records can be kept of accesses to the database, e.g. for security or accounting purposes. An active database system can be used to automatically keep these records in a system log table. This feature can easily be realized by means of transaction modification, as illustrated by the example below.

Example 8.5 Suppose we want to keep record of all insertions to the *trade* relation of the example beer database. Then a system log relation can be created with the following scheme:

$SysLog(user, time, event)$

Now the following rule implements the desired automatic logging feature:

WHEN $DEL(trade)$
IF NOT $false$
THEN $insert(SysLog, [\$user, \$time, "Data deleted from Trade"])$

In this rule, $\$user$ and $\$time$ denote system variables that contain the name of the current user and the system time, respectively. □

8.3 Constraint enforcement protocol extensions

In this thesis, integrity constraint enforcement is based on the transaction modification approach. This approach allows for an easy integration of integrity control into a real-world database management system. Further, parallelism in constraint enforcement has been shown to be an effective way to decrease the response time of modified transactions. In certain application environments with very high performance requirements, however, the transaction modification technique may not be efficient enough. This is mainly caused by the following two facts:

- Constraint enforcement through transaction modification always requires the transport of extended relational algebra constructs from data dictionary to transaction manager, and from transaction manager to the data management layer of the system. If many constraints are defined for which the evaluation is cheap, the relative overhead of this communication can be high.

- In the transaction modification approach to constraint enforcement, the data management layer of a system is completely unaware of integrity constraints. Therefore, no low level optimizations are possible to reduce the volume of data to be processed in constraint enforcement, or to avoid enforcement when a constraint cannot actually be violated.

These inefficiencies can be avoided by extending the constraint enforcement protocols. The following protocol classes can be distinguished [Gref90b]:

8. Extending the ideas

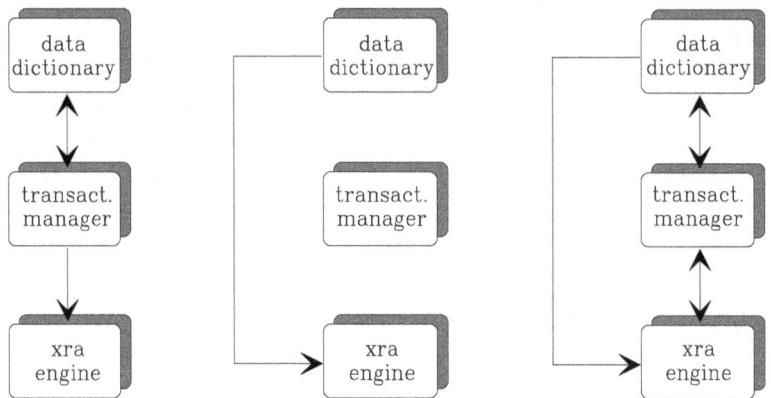

Figure 8.1: Explicit, implicit, and hybrid constraint enforcement

Explicit Enforcement In the explicit constraint enforcement protocol class, all constraint enforcement actions are explicitly controlled by the transaction management layer of the system. This is the situation used in the preceding chapters.

Implicit Enforcement In the implicit case, all constraint enforcement actions are implicitly performed by the data management layer, i.e. without any intervention of the transaction management layer. This protocol class aims at reducing the communication overhead in constraint enforcement.

Hybrid Enforcement In the hybrid enforcement protocol class, constraint enforcement is performed in co-operation by the transaction and data management layers of a system. This protocol class aims at low level optimizations in constraint enforcement to avoid unnecessary work.

The interfaces between data dictionary, transaction manager, and data management layer (extended relational algebra engine) are shown in Figure 8.1 for the three protocol classes. In the explicit enforcement case, the transaction manager retrieves the constraints from the data dictionary and sends commands to the processes in the relational engine; according to the transaction modification technique, this is performed at the end of a transaction. In the implicit enforcement case, the data dictionary directly informs the relational engine of integrity constraints; this can be performed at constraint definition time[2]. In the hybrid enforcement case, part of the constraint information is directed to the relational engine, and part follows the route through the transaction manager.

Explicit constraint enforcement has been discussed in detail in Chapters 4 to 6. Below, implicit and hybrid enforcement are discussed in short.

---

[2] If the constraint information is dependent on the fragmentation of the involved relation, the data dictionary will automatically inform the relational engine about the changes upon relation refragmentation.

8.3 Constraint enforcement protocol extensions

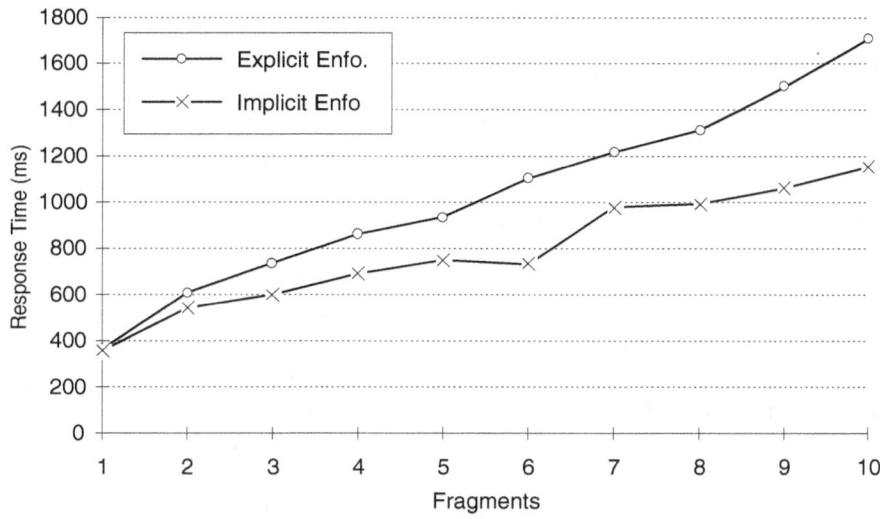

Figure 8.2: Explicit versus implicit enforcement response times

8.3.1 Implicit constraint enforcement

When implicit constraint enforcement is used, constraint enforcement is performed fully autonomously by the data management processes of a parallel database system. This implies a positive effect on constraint enforcement performance, but a negative effect on functionality.

As constraints are enforced autonomously by the data management processes, the communication overhead for constraint enforcement at transaction execution time is reduced to zero. This will have a positive effect on the response time of update transactions including integrity control. The gain increases with the number of data management processes involved in constraint enforcement. As PRISMA/DB is equipped with implicit constraint enforcement mechanisms, this can be illustrated by means of an experiment. Figure 8.2 shows the response times of the execution of a simple update transaction, including the enforcement of a domain constraint. Clearly, the use of implicit constraint enforcement performs better as the number of fragments increases.

The fact that no communication is necessary between transaction manager and data managers limits the types of constraints that can be handled by implicit constraint enforcement. As all communication between data managers is set up by the transaction manager, implicit enforcement can only be used for constraints that require no communication between data managers, i.e. constraints that can be fully enforced within the scope of one data manager (relation fragment). So, domain constraints and tuple constraints can be handled by implicit constraint enforcement, but referential integrity constraints cannot.

8.3.2 Hybrid constraint enforcement

When hybrid constraint enforcement is used, the transaction manager and data manager processes co-operate to reduce the cost of constraint enforcement. Two main forms of hybrid enforcement can be distinguished, one aiming at data reduction, the other at process reduction in global constraint enforcement. Both forms are discussed below and illustrated by means of an example.

Hybrid constraint enforcement can be used to locally, i.e. within the scope of a single data manager, reduce the volume of data to be processed in global, i.e. by multiple data managers, constraint enforcement. This can be achieved by the use of special-purpose differential sets for specific constraints[3]. The involved data managers are informed by the data dictionary about these differential sets. They use special-purpose algorithms to reduce the size of these differential sets during the execution of update transactions. In the constraint definitions, the standard differential sets are replaced by the special-purpose ones. As such, the normal transaction modification protocols can be used by the transaction manager.

Hybrid constraint enforcement can also be used to completely avoid global constraint enforcement in some cases. This requires the use of special-purpose differential sets plus the use of an extra communication protocol between transaction manager and data managers. Upon transaction execution, the data managers try to reduce the sizes of the special-purpose differential sets, aiming at empty sets. If the set is indeed empty, constraint enforcement can be skipped for the associated constraint. When the transaction manager receives the constraint definition from the data dictionary, it first communicates with the data managers to decide whether global constraint enforcement is actually needed for specific constraints. If not, it is simply skipped. If it is needed, the data reduction protocol sketched above is used.

Example 8.6 We take the beer database from Appendix B with referential constraint $I3$ as example here. If the database is fragmented, the following integrity program exists for a fragment $beer_k$ (see Chapter 5):

WHEN $INS(beer_k)$
THEN $copy(unique(\pi_{brewery} beer_k^+), t_1, \ldots, t_n)$;
$\quad\quad alarm((t_1 - \pi_{name} brewery_1) \cap \cdots \cap (t_n - \pi_{name} brewery_n)$

Now suppose the following transaction is executed:

$begin$
$insert(beer_k, new\_beers)$
end

---

[3] These differential sets can be seen as redundant data to support fast constraint enforcement. The use of redundant data for efficient integrity control has been described in [Bern80].

8.3 Constraint enforcement protocol extensions

$beer_k$	
Pilsener	Grand
Premium	Grand
Pilsener	Green Tree
Super Bock	Green Tree

new_beers	
Pale Ale	Grand
Stout	Green Tree
Pilsener	Beidwuser

Table 8.1: Beer tuples

Suppose further that $beer_k$ and new_beers contain the tuples shown in Table 8.1 (irrelevant attributes are left out). After the execution of the insert command, the standard differential set $beer_k^+$ contains 3 tuples, all of which have to be processed in explicit constraint enforcement. If hybrid enforcement with data reduction is used, the data manager managing $beer_k$ is informed of the fact that a foreign key differential set is to be maintained for attribute *brewery*. It can then reduce this special-purpose set to a single tuple, because the foreign keys of the other two tuples in the standard differential set are already present in the fragment. Consequently, global constraint enforcement has to process only one tuple. If hybrid enforcement with process reduction is used, the data manager tries to empty the special-purpose differential set (which would succeed if the Beidwuser tuple would not be inserted), and inform the transaction manager that global enforcement of the constraint is superfluous. □

As shown by the example above, hybrid enforcement can well be used for referential integrity constraints. It is also usable for other constraint types, like grouped aggregate constraints (if the relation is fragmented on the grouping attribute).

The effect of extended constraint enforcement protocols on transaction response times has to be investigated to determine their usefulness for overall system performance improvement. A constraint optimizer module in a system using these protocol extensions should be equipped with algorithms to choose between the various protocols in specific situations.

Chapter 9

Conclusions

This thesis describes concepts and techniques for integrity control in parallel database systems, based on the transaction modification approach. This section concludes the work with a number of general observations on the main topics of this thesis, transaction modification and parallelism, and the research on which this thesis is based. First, a short discussion of the transaction modification approach is presented, discussing its strong points and flaws. Next, conclusions are drawn about the use of parallelism for integrity control. The third section pays attention to the way theoretical and experimental research have been combined in the work presented in this thesis. Finally, the current status of the research is summarized and possibilities for future research are listed.

9.1 Transaction modification approach

The transaction modification approach to integrity control as described in this thesis is characterized by a number of properties that distinguish it from other approaches. In the first place, triggering of integrity constraint enforcement is based on syntactical analysis of a transaction. In the second place, constraint evaluation is fully integrated with normal transaction execution in an extended relational algebra context. In the third place, the conceptual description of the technique can easily be mapped to a highly modular system architecture. These characteristics determine the positive and negative qualities of the approach, to be discussed below.

In the transaction modification approach, the extended relational algebra is used as the enforcement vehicle for integrity constraints. The expressiveness of the algebra guarantees that the approach can be used for a broad range of constraint types. The fact that constraint enforcement is fully integrated with normal transaction execution enables a simple system architecture. There is no need for complex constraint or rule execution mechanisms as used in other approaches.

Clearly, this enables a relatively easy implementation of an integrity control subsystem with basic functionality, as used in the PRISMA/DB system. Further, this characteristic allows for an easy extension of the transaction modification technique to a parallel context with fragmented relations. The fact that constraint triggering is based on syntactical analysis, allows for parallelism between constraint enforcement control and execution in the case of complex transactions. Finally, the modular system architecture and simple semantics of the transaction modification technique enable simple extensions to both system functionality and system architecture as discussed in Chapter 8.

The syntax-based integrity rule triggering of the transaction modification approach can be seen as a worst-case technique, as rules are triggered whenever the database *may* have been modified in a way that can cause an integrity violation. This can lead to unnecessary constraint enforcement. Further, this may cause infinite triggering behaviour in some situations. Solutions for part of these problems have been discussed in Chapter 4. The fact that the extended relational algebra is used as constraint enforcement vehicle is one of the strong points of transaction modification, but also limits its functionality, as integrity violation response actions are also restricted to the extended relational algebra. Other rule-based approaches allow more freedom here. Finally, the fact that constraint enforcement is controlled by a single process per transaction can lead to control bottlenecks in constraint enforcement in highly parallel situations. Approaches that allow distributed control can be designed, but require more complex inter-process communication. The trade-off is not yet clear here.

9.2 Parallelism

Parallelism is the main line of approach in this thesis to deal with the high processing costs associated with integrity constraint enforcement. A number of conclusions can be drawn about the use of parallelism.

Parallelism has proven to be an efficient way to reduce the response times of transactions including integrity control. Experiments on PRISMA/DB have given satisfactory speedup results. As may be expected, the use of parallelism yields the best results on large volumes of data to be processed. The experiments have also shown, however, that the optimum degree of parallelism is limited by the relatively high control overhead necessary for transaction execution control. The overhead is strongly influenced by the type of the constraints and the fragmentation and allocation of the relations used in the transactions. The latter point indicates that data allocation algorithms should take integrity constraints into consideration.

Parallelism also adds to the problems to be solved in integrity control. In the first place, it must be taken into consideration in constraint optimization and translation. In the second place, the use of parallelism requires advanced action scheduling techniques for transaction execution, as discussed in Chapter 7 of this thesis. Finally, the use of parallelism complicates the experimental evaluation of the techniques, both because the number of parameters to be chosen for experiments is strongly increased and because the results are harder to interpret.

It is stressed again, however, that parallelism combines with transaction modification in both an elegant and an effective way. Other rule-based approaches to integrity control require more complex execution control mechanisms to deal with parallelism.

9.3 Theory and practice

As will be clear from the contents of the preceding chapters, the research underlying this thesis has been performed on the boundary between theory and practice, concepts and experiments. This fact has heavily influenced both the way the research has been conducted, and the results of the research.

The combination of theory and concepts on the one side, and practice and experiments on the other side, allows for a fruitful cross-fertilization. New theories and concepts can be implemented and tested in practice. This in turn can produce new insights that can be used to adjust or extend the theories and concepts. The practice of transaction modification in PRISMA/DB, for example, has lead to the insight that a clear conceptual background for dynamic action scheduling is necessary. Especially in a context of complex parallel systems, the combination of a conceptual and a practical approach is indispensable: leaving out one of the two ingredients may produce either a system that nobody can understand, or a system that nobody can use.

On the other hand, the combination places the research in a strong field of tension, both for pragmatic and conceptual reasons. The pragmatic reason may be clear, as the combination of two tasks that enlarge each other implies a large volume of work to be performed. The conceptual reason is the fact that some ideas may work fine in theory, but are very hard to use in practice, or the other way around, some ideas are quite usable in a system, but hard to conceptualize. Finding a good balance has been one of the major problems of this research, since restrictions on one or the other side are often easy to describe, but always hard to sell.

9.4 Current status and future research

The current status of the research on which this thesis is based can be summarized as follows. A complete conceptual framework for integrity control through transaction modification in a parallel database system has been developed. This framework links together a number of techniques necessary for the transaction modification approach. Part of these techniques is described in full detail, part requires further research. The concepts and algorithms developed have been applied in a simple prototype integrity control subsystem that is integrated in the parallel main-memory PRISMA/DB database system. Tests on this system have demonstrated the feasability of the transaction modification approach and the effectiveness of parallelism to reduce response times of modified transactions.

Future research in the field set out by this thesis can be conducted in three directions. The first research direction is the optimization and translation of integrity rules, i.e. the optimization of constructs in the $\mathcal{CL}$ constraint specification language and the translation of $\mathcal{CL}$ constructs to the extended relational algebra. The second direction leads to further development of the integrity control subsystem in the PRISMA/DB context and further investigation of the effects of parallelism on constraint enforcement, especially in the context of complex constraints and transactions. Finally, the third direction is concerned with extending the transaction modification ideas in one of the areas indicated in Chapter 8 of this thesis.

Bibliography

[Aho78]   A.V. Aho, J.D. Ullman; *Principles of Compiler Design*; Addison-Wesley, Reading, USA, 1978.

[Amer89] P. America; *Language Definition of POOL-X*; PRISMA Document P350; Philips Research Laboratories Eindhoven, The Netherlands, 1989.

[Amer90] P. America (ed.); *Proceedings of the PRISMA Workshop on Parallel Database Systems*; Noordwijk, The Netherlands, 1990; Springer-Verlag Lecture Notes in Computer Science 503, 1991.

[Apers88] P.M.G. Apers, M.L. Kersten, H.C.M. Oerlemans; *PRISMA Database Machine: A Distributed Main Memory Approach*; Proceedings of the International Conference on Extending Database Technology; Venice, Italy, 1988.

[Apers92a] P.M.G. Apers et al.; *PRISMA/DB: A Parallel Main-Memory Relational DBMS*; Memorandum INF92-12; University of Twente, The Netherlands, 1992;

[Apers92b] P.M.G. Apers et al.; *PRISMA/DB: A Parallel Main-Memory Relational DBMS*; To appear in IEEE Transactions on Knowledge and Data Engineering.

[Astr76] M.M. Astrahan et al.; *System R: A Relational Approach to Database Management*; ACM Transactions on Database Systems, Vol. 1, No. 2, June 1976.

[Auddi91] A. Auddino, E. Amiel, B. Bhargava; *Experiences with SUPER, a Database Visual Environment*; Proceedings 2nd of the International Conference on Database and Expert Systems Applications; Berlin, Germany, 1991.

[Badal79] D. Badal, G. Popek; *Cost and Performance Analysis of Semantic Integrity Validation Methods*; Proceedings of the 1979 ACM SIGMOD

International Conference on the Management of Data; Boston, USA, 1979.

[Beeri88]   C. Beeri, H.J. Schek, G. Weikum; *Multilevel Transaction Management, Theoretical Art or Practical Need?*; Proceedings of the 1988 International Conference on Extending Database Technology; Venice, Italy, 1988.

[Bern80]   P. Bernstein, B. Blaustein, E. Clarke; *Fast Maintenance of Semantic Integrity Assertions Using Redundant Aggregate Data*; Proceedings of the 6th International Conference on Very Large Data Bases; Montreal, Canada, 1980.

[Bitt83]   D. Bitton, D.J. DeWitt, C. Turbyfill; *Benchmarking Database Systems: A Systematic Approach*; Proceedings of the 9th International Conference on Very Large Data Bases; Florence, Italy, 1983.

[BMJ65]   BMJ Editorial; *Some Hospital Statistics*; British Medical Journal, Vol. 1, 1965.

[Boral90]   H. Boral et al.; *Prototyping Bubba, A Highly Parallel Database System*; IEEE Transactions on Knowledge and Data Engineering, Vol. 2, No. 1, 1990.

[Brat89]   K. Bratbergsengen, T. Gjelsvik; *The Development of the CROSS8 and HC16-186 (Database) Computers*; Proceedings of the 6th International Workshop on Database Machines; Deauville, France, 1989.

[Brod78]   M.L. Brodie; *Specification and Verification of Data Base Semantic Integrity*; Ph.D. Thesis, Dept. of Computer Science, University of Toronto; Toronto, Canada, 1978.

[Bry86]   F. Bry, R. Manthey; *Checking Consistency of Database Constraints: a Logical Basis*; Proceedings of the 12th International Conference on Very Large Data Bases; Kyoto, Japan, 1986.

[By91]   R.A. de By; *The Integration of Specification Aspects in Database Design*; Ph.D. Thesis; University of Twente, the Netherlands, 1991.

[Ceri84]   S. Ceri, G. Pelagatti; *Distributed Databases, Principles and Systems*; McGraw-Hill, New York, USA, 1984.

[Ceri90a]   S. Ceri, J. Widom; *Deriving Production Rules for Constraint Maintenance*; IBM Research Report RJ 7348; IBM Almaden Research Center, San Jose, USA, 1990.

[Ceri90b]   S. Ceri, J. Widom; *Deriving Production Rules for Constraint Maintenance*; Proceedings of the 16th International Conference on Very Large Data Bases; Brisbane, Australia, 1990.

[Ceri90c]   S. Ceri, G. Gottlob, L. Tanca; *Logic Programming and Databases*; Springer-Verlag, Berlin, Germany, 1990.

[Ceri91]   S. Ceri, J. Widom; *Deriving Production Rules for Incremental View Maintenance*; Proceedings of the 17th International Conference on Very Large Data Bases; Barcelona, Spain, 1991.

[Ceri92]   S. Ceri, J. Widom; *Production Rules in Parallel and Distributed Database Environments*; IBM Research Report RJ 8564; IBM Almaden Research Center, San Jose, USA, 1992.

[Cham76]   D.D. Chamberlin et al.; *SEQUEL 2: A Unified Approach to Data Definition, Manipulation, and Control*; IBM Journal of Research and Development, Vol. 20, No. 6, 1976.

[Cham81]   D.D. Chamberlin, A.M. Gilbert, R.A. Yost; *A History of System R and SQL/Data System*; Proceedings of the 7th International Conference on Very Large Data Bases; Cannes, France, 1981.

[Chom92]   J. Chomicki; *History-less Checking of Dynamic Integrity Constraints*; Proceedings of the 8th IEEE International Conference on Data Engineering; Phoenix, USA, 1992.

[Codd70]   E.F. Codd; *A Relational Model for Large Shared Data Banks*; Communications of the ACM, Vol.13, No.6, 1970.

[Codd71]   E.F. Codd; *A Data Base Sublanguage Founded on Relational Calculus*; Proceedings of the 1971 ACM SIGFIDET Workshop; 1971.

[Codd79]   E.F. Codd; *Extending the Database Relational Model to Capture More Meaning*; ACM Transactions on Database Systems, Vol. 4, No. 4, 1979.

[Crem83]   A.B. Cremers, G. Domann; *AIM - An Integrity Monitor for the Database System INGRES*; Proceedings of the 9th International Conference on Very Large Data Bases; Florence, Italy, 1983.

[Date81]   C.J. Date; *Referential Integrity*; Proceedings of the 7th International Conference on Very Large Data Bases; Cannes, France, 1981.

[Date83]   C.J. Date; *An Introduction to Database Systems, Volume II*; Addison Wesley, Reading, USA, 1983.

[Date90a]   C.J. Date; *Referential Integrity and Foreign Keys: Further Considerations*; in *Relational Database Writings 1985-1989*; Addison-Wesley, Reading, USA, 1990.

[Date90b] C.J. Date; *A Contribution to the Study of Database Integrity*; in *Relational Database Writings 1985-1989*; Addison-Wesley, Reading, USA, 1990.

[IBM89] *IBM DATABASE 2 Referential Integrity Usage Guide*; IBM Corporation, 1989.

[DeWi90] D.J. DeWitt et al.; *The GAMMA Database Machine Project*; IEEE Transactions on Knowledge and Data Engineering; Vol. 2, No. 1, 1990.

[Ehri84] H.D. Ehrich, U.W. Lipeck, M. Gogolla; *Specification, Semantics and Enforcement of Dynamic Database Constraints*; Proceedings of the 10th International Conference on Very Large Data Bases; Singapore, 1984.

[Eich88] M.H. Eich, D.L. Wells; *Database Concurrency Control Using Data Flow Graphs*; ACM Transactions on Database Systems, Vol.13, No.2, 1988.

[Elmas89] R. Elmasri, S.B. Navathe; *Fundamentals of Database Systems*; Benjamin/Cummings, Redwood, USA, 1989.

[Eswa75] K.P. Eswaran, D.D. Chamberlin; *Functional Specifications of a Subsystem for Data Base Integrity*; Proceedings of the 1st International Conference on Very Large Data Bases; Framingham, USA, 1975.

[Gard79] G. Gardarin, M. Melkanoff; *Proving Consistency of Database Transactions*; Proceedings of the 5th International Conference on Very Large Data Bases; Rio de Janeiro, Brazil, 1979.

[Gard83] G. Gardarin et al.; *Design of a Multiprocessor Relational Database System*; IFIP 9th World Computer Congress; Paris, France, 1983.

[Gard89] G. Gardarin, P. Valduriez; *Relational Databases and Knowledge Bases*; Addison-Wesley, Reading, USA, 1989.

[Gill76] A.Gill; *Applied Algebra for the Computer Sciences*; Prentice-Hall, Englewood Cliffs, USA, 1976.

[Gray81] J. Gray; *The Transaction Concept: Virtues and Limitations*; Proceedings of the 7th Conference on Very Large Data Bases; Cannes, France, 1981.

[Gref88] P.W.P.J. Grefen, A.N. Wilschut, P.M.G. Apers, M.L. Kersten; *Implementing PRISMA/DB in an OOPL*; Memorandum INF88-69; University of Twente, The Netherlands, 1988.

[Gref89a]   P.W.P.J. Grefen; *Integrity Constraint Handling in a Parallel Database System*; Memorandum INF89-59; University of Twente, The Netherlands, 1989.

[Gref89b]   P.W.P.J. Grefen; *User Manual PRISMA/DB0*; PRISMA Project Document P415; University of Twente, The Netherlands, 1989.

[Gref90a]   P.W.P.J. Grefen, P.M.G. Apers; *Parallel Handling of Integrity Constraints on Fragmented Relations*; Proceedings of the International Symposium on Databases in Parallel and Distributed Systems; Dublin, Ireland, 1990.

[Gref90b]   P.W.P.J. Grefen, J. Flokstra, P.M.G. Apers; *Parallel Handling of Integrity Constraints*; Proceedings of the PRISMA Workshop on Parallel Database Systems; Noordwijk, The Netherlands, 1990.

[Gref90c]   P.W.P.J. Grefen; *Integrity Constraint Enforcement through Transaction Modification*; Memorandum INF90-60; University of Twente, The Netherlands, 1990.

[Gref90d]   P.W.P.J. Grefen; *A Survey into Integrity Constraints*; PRISMA Project Document P246; University of Twente, The Netherlands, 1990.

[Gref90e]   P.W.P.J. Grefen; *Design Considerations for Integrity Constraint Handling in PRISMA/DB1*; PRISMA Project Document P508; University of Twente, The Netherlands, 1990.

[Gref91a]   P.W.P.J. Grefen, A.N. Wilschut, J. Flokstra; *PRISMA/DB 1.0 User Manual*; Memorandum INF91-06; University of Twente, The Netherlands, 1991.

[Gref91b]   P.W.P.J. Grefen, P.M.G. Apers; *Integrity Constraint Enforcement through Transaction Modification*; Proceedings of the 2nd International Conference on Database and Expert Systems Applications; Berlin, Germany, 1991.

[Gref91c]   P.W.P.J. Grefen; *Dynamic Action Scheduling in a Parallel Database System*; Memorandum INF91-58; University of Twente, The Netherlands, 1991;

[Gref91d]   P.W.P.J. Grefen; *Integrity Control in Relational Database Systems - An Overview*; Memorandum INF91-80; University of Twente, The Netherlands, 1991.

[Gref92a]   P.W.P.J. Grefen, P.M.G. Apers; *Dynamic Action Scheduling in a Parallel Database System*; Proceedings of the 4th International Conference on Parallel Architectures and Languages Europe 1992; Paris, France, 1992;

[Gref92b] P.W.P.J. Grefen, J. Flokstra, P.M.G. Apers; *Performance Evaluation of Constraint Enforcement in a Parallel Main-Memory Database System*; Proceedings of the 3rd International Conference on Database and Expert Systems Applications; Valencia, Spain, 1992.

[Haas90] L. Haas et al.; *Starburst Midflight: As the Dust Clears*; IEEE Transactions on Knowledge and Data Engineering, Vol 2., No. 1, 1990.

[Hamm75] M.M. Hammer, D.J. McLeod; *Semantic Integrity in a Relational Data Base System*; Proceedings of the 1st International Conference on Very Large Data Bases; Framingham, USA, 1975.

[Hart88] B.E. Hart, S. Danforth, P. Valduriez; *Parallelizing a Database Programming Language*; Proceedings of the International Symposium on Databases in Parallel and Distributed Systems; Austin, Texas, USA, 1988.

[Held75] G. Held et al.; *INGRES: A Relational Data Base System*; Proceedings of the 1975 National Computer Conference; Anaheim, USA, 1975.

[Hsu85] A. Hsu, T. Imielinsky; *Integrity Checking for Multiple Updates*; Proceedings of the 1985 ACM SIGMOD International Conference on the Management of Data; Austin, USA, 1985.

[Jack77] M. Jackson; *The World Guide to Beer*; Quarto Publishing Limited, London, UK, 1977.

[Kers87] M.L. Kersten et al.; *A Distributed Main Memory Database Machine*; Proceedings of the 5th International Workshop on Database Machines; Karuizawa, Japan, 1987.

[Korth86] H.F. Korth, A. Silberschatz; *Database System Concepts*; McGraw-Hill, New York, USA, 1986.

[Li88] K. Li, J.F. Naughton; *Multiprocessor Main Memory Transaction Processing*; Proceedings of the International Symposium on Databases in Parallel and Distributed Systems; Austin, Texas, USA, 1988.

[Lohm91] G.M. Lohman, B. Lindsay, H. Pirahesh, K.B. Schiefer; *Extensions to Starburst: Objects, Types, Functions, and Rules*; Communications of the ACM, Vol.34, No.10, 1991.

[Mark91] V.M. Markowitz; *Safe Referential Integrity Structures in Relational Databases*; Proceedings of the 17th International Conference on Very Large Data Bases; Barcelona, Spain, 1991.

[Morg84] M. Morgenstern; *Constraint Equations: Declarative Expression of Constraints With Automatic Enforcement*; Proceedings of the 10th International Conference on Very Large Data Bases; Singapore, 1984.

[Motro89]   A. Motro; *Integrity = Validity + Completeness*; ACM Transactions on Database Systems, Vol. 14, No. 4, 1989.

[Nico78]   J.M. Nicolas; *First Order Logic Formalization for Functional, Multivalued, and Mutual Dependencies*; Proceedings of the 1978 ACM SIGMOD International Conference on Management of Data; Austin, USA, 1978.

[Nico82]   J.M. Nicolas; *Logic for Improving Integrity Checking in Relational Data Bases*; Acta Informatica, Vol. 18, 1982.

[Noble86]   H. Noble, T. Abbod; *Meta-Rules and Semantic Integrity Constraints in Databases*; Proceedings of the 5th British National Conference on Databases; Canterbury, UK, 1986.

[Pare89]   J. Paredaens, P. de Bra, M. Gyssens, D. van Gucht; *The Structure of the Relational Database Model*; Springer-Verlag, Berlin, Germany, 1989.

[Qian86]   X. Qian, G. Wiederhold; *Knowledge-based Integrity Constraint Validation*; Proceedings of the 12th International Conference on Very Large Data Bases; Kyoto, Japan, 1986.

[Raju88]   K.V.S.V.N. Raju, A.K. Majumdar; *Fuzzy Functional Dependencies and Lossless Join Decomposition of Fuzzy Relational Database Systems*; ACM Transactions on Database Systems, Vol. 13, No. 2, 1988.

[Rowe87]   L. Rowe, M. Stonebraker; *The POSTGRES Data Model*; Proceedings of the 13th International Conference on Very Large Data Bases; Brighton, UK, 1987.

[Satoh85]   K. Satoh, M. Tsuchida, F. Nakamure, K. Oomachi; *Local and Global Query Optimization Mechanisms for Relational Databases*; Proceedings of the 11th International Conference on Very Large Databases; Stockholm, Sweden, 1985.

[Schön89]   H. Schöning; *Preserving Consistency in Nested Transactions*; Department of Computer Science; University of Kaiserslautern, Germany, 1989.

[Simon84]   E. Simon, P. Valduriez; *Design and Implementation of an Extendible Integrity Subsystem*; Proceedings of the 1984 ACM SIGMOD Annual Meeting; Boston, USA, 1984.

[Simon85]   E. Simon, P. Valduriez; *Integrity Control in Distributed Database Systems*; MCC Technical Report DB-103-85; MCC, Austin, USA, 1985.

[Simon87]   E. Simon, P. Valduriez; *Design and Analysis of a Relational Integrity Subsystem*; MCC Technical Report DB-015-87; MCC, Austin, USA, 1987.

[Skelt92]   C.J. Skelton et al.; *EDS: A Parallel Computer System for Advanced Information Processing*; Proceedings of the 4th International Conference on Parallel Architectures and Languages Europe; Paris, France, 1992.

[Small86]   C. Small; *An Implementation of a Constraint Enforcement System*; Proceedings of the 5th British National Conference on Databases; Canterbury, UK, 1986.

[Spek90]   J. v.d. Spek; *POOL-X and its Implementation*; Proceedings of the PRISMA Workshop on Parallel Database Systems; Noordwijk, The Netherlands, 1990.

[Stone75]   M. Stonebraker; *Implementation of Integrity Constraints and Views by Query Modification*; Proceedings of the 1975 ACM SIGMOD International Conference on the Management of Data; San Jose, USA, 1975.

[Stone86a]   M. Stonebraker (ed.); *The INGRES Papers*; Addison-Wesley, Reading, USA, 1986.

[Stone86b]   M. Stonebraker, L. Rowe; *The Design of POSTGRES*; Proceedings of the 1986 ACM SIGMOD International Conference on Management of Data; Washington DC, USA, 1986.

[Stone88]   M. Stonebraker, E.N. Hanson, S. Potamianos; *The POSTGRES Rule Manager*; IEEE Transactions on Software Engineering, Vol. 14, No. 7, 1988.

[Stone90a]   M. Stonebraker, L.A. Rowe, M. Hirohama; *The Implementation of POSTGRES*; IEEE Transactions on Knowledge and Data Engineering, Vol. 2, No. 1, 1990.

[Stone90b]   M. Stonebraker, A. Jhingran, J. Goh, S. Potamianos; *On Rules, Procedures, Caching and Views in Data Base Systems*; Proceedings of the 1990 ACM SIGMOD International Conference on Management of Data; Atlantic City, USA, 1990.

[Su88]   S.Y.W. Su; *Database Computers: Principles, Architectures, and Techniques*; McGraw-Hill, New York, USA, 1988.

[Super90]   *SuperBase User and Reference Manual*; Precision Software Limited, Irving, USA, 1990.

[Tsich82] D.C. Tsichritzis, F.H. Lochovsky; *Data Models*; Prentice-Hall, Englewood Cliffs, USA, 1982.

[Ullm82] J.D. Ullman; *Principles of Database Systems, Second Edition*; Computer Science Press, Rockville, USA, 1982.

[Vald89] P. Valduriez, G. Gardarin; *Analysis and Comparison of Relational Database Systems*; Addison-Wesley, Reading, USA, 1989.

[Vlot90] M.C. Vlot; *The POOMA Architecture*; Proceedings of the PRISMA Workshop on Parallel Database Systems; Noordwijk, The Netherlands, 1990.

[Wang91] X.Y. Wang, N.J. Fiddian, W.A. Gray; *The Development of a Knowledge-Based Database Transaction Design Assistant*; Proceedings of the 2nd International Conference on Database and Expert Systems Applications; Berlin, Germany, 1991.

[Wats90] P. Watson, P. Townsend; *The EDS Parallel Relational Database System*; Proceedings of the PRISMA Workshop on Parallel Database Systems; Noordwijk, The Netherlands, 1990.

[Widom90] J. Widom, S.J. Finkelstein; *Set-Oriented Production Rules in Relational Database Systems*; Proceedings of the 1990 ACM SIGMOD International Conference on Management of Data; Atlantic City, USA, 1990.

[Widom91] J. Widom, R.J. Cochrane, B.G. Lindsay; *Implementing Set-Oriented Production Rules as an Extension to Starburst*; Proceedings of the 17th International Conference on Very Large Data Bases; Barcelona, Spain, 1991.

[Wils89] A.N. Wilschut, P.W.P.J. Grefen, P.M.G. Apers, M.L. Kersten; *Implementing PRISMA/DB in an OOPL*; Proceedings of the 6th International Workshop on Database Machines; Deauville, France, 1989.

[Wils90] A.N. Wilschut, P.M.G. Apers; *Pipelining in Query Execution*; Proceedings of the International Conference on Databases, Parallel Architectures, and Their Applications; Miami Beach, Florida, USA, 1990.

[Wils91] A.N. Wilschut, P.M.G. Apers; *Dataflow Query Execution in a Parallel Main-Memory Environment*; Proceedings of the 1st International Conference on Parallel and Distributed Information Systems; Miami Beach, Florida, USA, 1991.

[Zloof75] M.M. Zloof; *Query-By-Example: The Invocation and Definition of Tables and Forms*; Proceedings of the 1st International Conference on Very Large Data Bases; Framingham, USA, 1975.

[Zloof78]   M.M. Zloof; *Security and Integrity within the Query-by-Example Database Management Language*; IBM Research Report RC6982; IBM Thomas J. Watson Research Center, Yorktown Heights, USA, 1978.

Appendix A

Extended relational algebra summary

This appendix gives an overview of the syntax of the extended relational algebra developed in this thesis. The basis for the language is defined in Chapter 2, Definitions 2.6 to 2.11 and 2.14. Extensions are described in Definitions 4.1, 5.8, and 8.2. The algebra is a variant of the XRA language used in the PRISMA/DB system; a complete description of XRA can be found in [Gref91a].

Syntax

In the syntax overview below, non-terminal symbols are denoted in brackets, like $\langle non\text{-}terminal \rangle$; terminal symbols are printed in bold face, like **terminal**.

Transaction

$\langle transaction \rangle$ ::= **begin**
$\langle program \rangle$
end

$\langle program \rangle$ ::= $\mathbf{P}_\varepsilon |$
$\langle statement \rangle |$
$\langle statement \rangle ; \langle program \rangle$

Statement

$\langle statement \rangle$ ::= $\langle assignment \rangle |$
$\langle dml\_stat \rangle |$
$\langle query\_stat \rangle |$
$\langle alarm\_stat \rangle$

$\langle assignment \rangle$ ::= $\langle simple\_ass \rangle |$
$\langle split\_stat \rangle |$
$\langle copy\_stat \rangle |$

$\langle simple\_ass \rangle$ ::= $\langle variable \rangle = \langle expression \rangle$

$\langle split\_stat \rangle$ ::= **split**($\langle expression \rangle, \langle attributes \rangle, \langle var\_list \rangle$)

$\langle copy\_stat \rangle$ ::= **copy**($\langle expression \rangle, \langle var\_list \rangle$)

$\langle dml\_stat \rangle$ ::= $\langle insert\_stat \rangle |$
$\langle delete\_stat \rangle |$
$\langle update\_stat \rangle$

$\langle insert\_stat \rangle$ ::= **insert**($\langle variable \rangle, \langle expression \rangle$)

$\langle delete\_stat \rangle$ ::= **delete**($\langle variable \rangle, \langle expression \rangle$)

$\langle update\_stat \rangle$ ::= **update**($\langle variable \rangle, \langle expression \rangle, \langle attr\_exprs \rangle$)

$\langle query\_stat \rangle$ ::= ?$\langle expression \rangle$

$\langle alarm\_stat \rangle$ ::= **alarm**($\langle expression \rangle$)

Expression

$\langle expression \rangle$ ::= $\langle variable \rangle |$
$\langle select \rangle |$
$\langle project \rangle |$
$\langle unique \rangle |$
$\langle groupby \rangle |$
$\langle closure \rangle |$
$\langle product \rangle |$
$\langle join \rangle |$
$\langle union \rangle |$
$\langle difference \rangle |$
$\langle intersect \rangle$

$\langle select \rangle$ ::= $\sigma_{\langle condition \rangle} \langle expression \rangle$

$\langle project \rangle$ ::= $\pi_{\langle attr\_exprs \rangle} \langle expression \rangle$

$\langle unique \rangle$ ::= **unique**($\langle expression \rangle$)

$\langle groupby \rangle$ ::= **groupby**
($\langle attributes \rangle, \langle aggregate \rangle, \langle attribute \rangle, \langle expression \rangle$)

$\langle closure \rangle$ ::= **closure**($\langle attributes \rangle, \langle condition \rangle, \langle expression \rangle$)

$\langle product \rangle$ ::= $\langle expression \rangle \times \langle expression \rangle$

$\langle join \rangle$ ::= $\langle expression \rangle \bowtie_{\langle condition \rangle} \langle expression \rangle$

$\langle union \rangle$ ::= $\langle expression \rangle \cup \langle expression \rangle$

$\langle difference \rangle$ ::= $\langle expression \rangle - \langle expression \rangle$

$\langle intersect \rangle$ ::= $\langle expression \rangle \cap \langle expression \rangle$

Attributes

$\langle attributes \rangle$ ::= ($\langle attr\_list \rangle$)

$\langle attr\_list \rangle$ ::= $\langle attribute \rangle |$
$\langle attribute \rangle, \langle attr\_list \rangle$

$\langle attr\_exprs \rangle$ ::= ($\langle aexpr\_list \rangle$)

$\langle aexpr\_list \rangle$ ::= $\langle attr\_expr \rangle |$
$\langle attr\_expr \rangle, \langle aexpr\_list \rangle$

$\langle attribute \rangle$ ::= %$\langle integer \rangle |$
$\langle identifier \rangle$

Various

$\langle var\_list \rangle$ ::= $\langle variable \rangle |$
$\langle variable \rangle, \langle var\_list \rangle$

$\langle variable \rangle$ ::= $\langle identifier \rangle$

$\langle aggregate \rangle$ ::= **CNT|SUM|AVG|MIN|MAX**

Appendix B
Example database

This appendix describes the example database schema and integrity constraints defined on it that are used throughout this thesis to illustrate the various techniques. The database is a simple beer database, modeling a universe of discourse dealing with beer brewing and drinking. Note that the example beer data used in this thesis are fictive; for real-world information about beer, the reader is kindly referred to a standard work like [Jack77].

The database schema of the example beer database is shown in Table B.1; it consists of the following relations:

1. A relation *beer* describing of each beer the name, brewery, type and alcohol percentage.

2. A relation *brewery* describing of each brewery the name, location, and the concern the brewery belongs to.

3. A relation *pub* describing of each pub the name, the location, and the brewery the pub is affiliated to, if any.

4. A relation *trade* describing per combination of pub and beer the quantity of the beer bought and sold by the pub.

The example constraints are shown in Table B.2 in first order logic notation. They are the following:

1. Constraint $I1$ is a simple domain constraint on an attribute of relation *beer*. It states that a beer cannot contain less than 0% alcohol.

2. Constraint $I2$ is a uniqueness (key) constraint, stating that the name of a brewery must be unique.

3. Constraints $I3$ and $I4$ are referential integrity constraints from relation *beer* to *brewery* respectively from *trade* to *beer*. The constraints specify that every

relation	attributes
beer	name, brewery, type, alcperc
brewery	name, city, country, part_of
pub	name, city, country, brewery
trade	pub, beer, qty_bought, qty_sold

Table B.1: Relations of example database

name	definition
$I1$	$(\forall x)(x \in beer \Rightarrow x.alcperc \geq 0)$
$I2$	$(MLT(brewery \leq 1)) \land$
	$(\forall x, y)((x, y \in brewery \land x.name = y.name) \Rightarrow (x = y))$
$I3$	$(\forall x)(x \in beer \Rightarrow (\exists y)(y \in brewery \land x.brewery = y.name))$
$I4$	$(\forall x)(x \in trade \Rightarrow (\exists y)(y \in beer \land x.beer = y.name))$
$I5$	$(\forall x)(x \in trade \Rightarrow x.qty\_bought \geq x.qty\_sold)$
$I6$	$AVG(beer, alcperc) > 3$
$I7$	$(\forall x, y)((x \in beer \land y \in beer_{old} \land x.name = y.name) \Rightarrow$
	$(x.alcperc \geq y.alcperc))$

Table B.2: Constraints of example database

beer must be brewed by an existing brewery, and that every sold beer must exist.

4. Constraint $I5$ is a tuple constraint on relation *trade*, stating that every pub must buy at least as much of each beer as it sells of the same beer.

5. Constraint $I6$ is an aggregate constraint on relation *beer* stating that the average alcohol percentage of all beers should exceed 3%.

6. Constraint $I7$ is a transition constraint on relation *beer*, using virtual relation $beer_{old}$, and stating that a new version of a beer should at least be as strong as the old version.

Index

alarm statement, 45
allocation, 62
    schema, 63
    transparency, 64
architecture
    action scheduler, 115
    integrity control
        abstract, 56
        in PRISMA/DB, 87
    PRISMA/DB
        basic, 83
        with integrity control, 87
    scheduling
        centralized, 116
        distributed, 116
atomicity, 13

concurrency control, 112
constraint, 15
    characteristics, 23
    compiler, 87
    extended, 126
        aggregate, 126
        recursive, 127
    optimization, 43, 71
    specification, 25
    state, 15
    transition, 15
    translation, 46, 74
copy statement, 74

data
    allocation, *see* allocation
    fragmentation, *see* fragmentation
database, 7

active, 129
    distributed, 63
    instance, 7
    schema, 7
    state, 7
        intermediate, 12
    transition, 7
    universe, 7
dependency
    direct order, 104
    direct resource, 111
    order, 104
    resource, 111
DisFrRS algorithm, 70
DistR algorithm, 69
domain, 6
durability, 13

fragmentation, 62
    schema, 62
    transparency, 64
FragR algorithm, 66
FragS algorithm, 67

G-graph, 113
GenTrigC algorithm, 41
GetIntP algorithm, 55
GetTrigP algorithm, 49
GetTrigPX algorithm, 54

integrity
    constraint, *see* constraint
    rule, *see* rule
    violation
        detection, 34

handling, 19
    prevention, 34

ModT algorithm, 51, 71

O-graph, 106
OptR algorithm, 43

parallelism, 60, 76
    in PRISMA/DB, 85
performance, 60, 82, 101, 131
    benchmark, 91
    evaluation, 90
    measurements, 93, 121, 133
POOL-X, 81
POOMA, 80
PRISMA, 79
PRISMA/DB, 82
    architecture, 83
    parallelism, 85
protocol extension, 131
    hybrid, 134
    implicit, 133

query
    modification, 35
    rewrite, 52

R-graph, 112
relation, 6
    differential, 43
    distributed, 63
    fragment, 62
    instance, 6
    schema, 6
relational algebra
    basic, 7
    extended, 9, 151
        aggregate functions, 9
        expression, 10
        program, 11
        statement, 11
        transaction, 12
    standard, 8
rule, 22

fragmentation, 64
    canonical, 66
    distributed, 68
    optimization, 43, 54
    selection, 49
    specification, 28, 38
    translation, 44, 54
    validation, 39

scheduling, 101
    global, 114
    O-graph based, 109
SelRS algorithm, 50
serializability, 13
SimpC algorithm, 69
split statement, 74

transaction, 12
    correct, 16
    modification, 35, 51, 70
    modifier, 51
    optimization, 106
    safe, 17
    workspace, 35
TransC algorithm, 46
TransR algorithm, 45
trigger set, 22
    generation, 41
TrOptRS algorithm, 48
tuple, 6

Curriculum Vitae

Paul Grefen was born in Heerlen, Limburg, on April 18th 1963. In 1975 he started his secondary education at the Gymnasium of St. Bernardinus College in Heerlen. In 1981 he passed the final exams cum laude.

Thereafter, he started his study Computer Science at the University of Twente. During his study he was a committee member of the Computer Science Study Society for several years. In 1986 he obtained a M.Sc. degree in the field of information system design. In the same year he was appointed as assistant professor at the Department of Computer Science at the University of Twente, where he worked as faculty education program coordinator.

In 1987 he joined the PRISMA project as a full-time researcher, with research emphasis on transaction management in parallel database systems. In the context of this project he started research on parallel integrity control in early 1989. This research has led to several international publications and a prototype implementation of a parallel integrity control subsystem.

In 1992 he switched back to an assistant professor position in the Department of Computer Science at the University of Twente, where he added groupware systems to his field of interest.

www.ingramcontent.com/pod-product-compliance
Lightning Source LLC
Chambersburg PA
CBHW082204220526

45470CB00010B/3040